Rebeca Figueira
Flávio Duarte

The use of information systems by the generations

Rebeca Figueira
Flávio Duarte

The use of information systems by the generations

Baby Bommers, X, Y and Z

ScienciaScripts

Imprint
Any brand names and product names mentioned in this book are subject to trademark, brand or patent protection and are trademarks or registered trademarks of their respective holders. The use of brand names, product names, common names, trade names, product descriptions etc. even without a particular marking in this work is in no way to be construed to mean that such names may be regarded as unrestricted in respect of trademark and brand protection legislation and could thus be used by anyone.

Cover image: www.ingimage.com

This book is a translation from the original published under ISBN 978-613-9-62543-7.

Publisher:
Sciencia Scripts
is a trademark of
Dodo Books Indian Ocean Ltd. and OmniScriptum S.R.L publishing group

120 High Road, East Finchley, London, N2 9ED, United Kingdom
Str. Armeneasca 28/1, office 1, Chisinau MD-2012, Republic of Moldova, Europe
Printed at: see last page
ISBN: 978-620-7-73044-5

Copyright © Rebeca Figueira, Flávio Duarte
Copyright © 2024 Dodo Books Indian Ocean Ltd. and OmniScriptum S.R.L publishing group

SUMMARY

ACKNOWLEDGMENTS

I would like to thank Professor Rebeca Formiga who gave me this unique opportunity to develop in the world of information systems, my family who have always been by my side, especially my mother Risomar and my father Antônio who taught me a lot in my studies, and my friends.

To all those who directly or indirectly contributed to the construction of this study and participated in this important stage of my life. I would also like to thank all the teachers who played a part in my education.

SUMMARY

This research aims to analyze the evolution of the use of information systems by the generations: *Baby Boomers*, X, Y and Z, with the specifications of verifying the evolution of the information systems available in each generation, analyzing the development of Information Technology and identifying the contributions and use of Information Technology by the different generations over time. This study is classified as quantitative, and the information collected was measured by means of questionnaires, in which the different characteristics of the generations and between them were observed, as well as their attitudes towards the evolution of information technologies and systems, in terms of their resistance to change, their use and their adaptations. The results obtained show that the majority of generations use information systems because of their everyday practicality at work or in their personal lives and that the *Boomers* are more resistant to technological change, Generation X are more eager to learn technologies and Generation Y and Z have shown that they are more accustomed to and not afraid of new technological experiences.

Keywords: Evolution; Information System; Technology; Generations.

1 INTRODUCTION

The history of computing begins a few years BC, when ancient civilizations used a wooden structure with rings that had symbols representing mathematical calculations and logical reasoning, to calculate quantities or sales in the commerce of the time. Between 1935 and 1938, the first computer was created with the aim of computing strategic calculations for the Second World War and, consequently, advances in the field of technology (MESSINA, 2017). Since then, civilizations have improved over the years, until the then coming Age of Technology.

Computers are being used more and more both personally and in organizations. Its users are of various ages, from children to adults and even senior citizens. However, there are still those who shy away from using computers. For this reason, this study aims to show the evolution of information systems as a development tool in the lives and daily lives of different generations.

The technological development observed in recent decades has been transforming the behavior of people and companies, with its environmental variables. In the micro-environment, for example, Information Technology (IT) is increasingly influencing people's daily lives. On the one hand, this makes it easier for consumers to shop online, communicate, entertain and enjoy themselves, and on the other, it makes it easier for the various business sectors to communicate with each other. In addition, IT also influences the macro-environment of society, i.e. the relationships between companies and all their *stakeholders*.

The Baby Boomers, X, Y and Z generations emerged after the Second World War, with the advance of technology and the growth in the human birth rate that occurred from the 1940s onwards (SERRANO, 2010). As a result, each generation experienced the advance of technology in a different way. The *Boomers*, for example, with the arrival of the computer and color television (TV). Generation X, on the other hand, saw technological advances with the use of the personal computer. Generation Y, in turn, had to adapt to the mobile phone and finally, Generation Z, which is considered to be the current generation, also known as *Millennials,* has to constantly deal with the variety of portable technologies such as *Smartphones*, *tablets and* others. From this perspective, it can be understood that technological advances and generations have a direct relationship with people's characteristics.

According to a survey of 18,000 professionals and students in 19 countries by researchers

Bresman and Rao (2017), in an article published in the *Harvard Business Review,* it was found that generations are different. It is known that each one has its own singularities, different beliefs, customs and cultures of the time. These particularities end up influencing consumption habits and consumer characteristics, especially when it comes to information systems.

The evolution of technology has come about as a result of the human need for information, for industrial machinery to produce their products, to obtain knowledge and transmit it quickly and effectively, to be able to study, explore and preserve our natural resources, for example, the need to communicate with other people over long distances, to make strategic calculations, to store information without piles of paper, to be able to carry research, budgets, spreadsheets and documents in a simple little object called a *Pen Drive*. Technology has become an essential part of people's lives and every day there are new developments that bring benefits to people's daily lives, whether in professional or personal activities.

In statistics published by BBC Brasil in 2015, IBGE reported that 85.6 million Brazilians over the age of 10 (49.4% of the population) had used the internet at least once in 2013. In addition, 78.3 million of those interviewed (45.3% of the population) said they had accessed the web via a computer and the rest used other types of equipment such as *tablets* and cell phones.

Over the decades since 1935, when the first generation appeared and the arrival of the computer, the world has evolved more and more until we reach the present day where we find high-tech resources and the tendency is to evolve even more, with which other generations will emerge. This research studied the history of the generations up to 2017. To this end, in order to analyze the evolution obtained, it sought to solve the following research question: What is the perceived evolution in the use of information systems by the *Baby Boomers,* X, Y and Z generations? This question will be answered in line with the specific objectives of this work.

1.1 OBJECTIVES

1.1.1 General objective

The main objective of this study is to analyze the evolution in the use of information systems and resistance to change by the *Baby Boomer*, X, Y and Z generations.

To this end, it was necessary to solve the specific objectives defined below.

1.1.2 Specific objectives

- Check the evolution of the information systems available in each generation.
- Analyze the development of Information Technology and the adaptation of generations in its use over time.
- Identify the behavior of the different generations and their contributions to the use of information systems.

1.2 BACKGROUND

The aim of this research was to analyze the behavior of the generations in relation to the use of information systems and their evolution over the years. In addition, to identify whether the generations that grew up with the advance of technology have managed to adapt to these major changes and also whether the current generation feels any difficulty in using new technologies.

From this perspective, this study is important because it attempts to verify consumer behavior through the evolution of technology, perceiving its difficulties. In this way, this research can be explored on a specific professional occasion where it can deal with generations at work and possibly with customers and suppliers, analyzing how they use information systems and so that the company's management can develop the necessary training for internal employees. This will generate continuous improvement in the execution of the company's activities.

In the administrative area of companies, research can help in internal and external selection decisions, where the profile of an employee can be selected for a particular sector that requires more knowledge in handling technology. Companies that offer products online for purchase can choose a particular product to market on the site according to the profile of the end consumer or build a site with handling that facilitates online purchases. For these reasons, it is clear that this is a timely and important topic for academia and companies in the region.

In addition, society will be able to use this work to understand the behavior of the generations and how they adapted to the evolution of technology. This will provide more information on how the first information systems, such as the computer, television and radio, came about, generating a more grounded understanding of the names and reasons why the *Baby Boomer,* X, Y and Z generations exist. In this way, the community will be able to understand the profiles of their family and friends in relation to technology, from children to the elderly.

2 RESEARCH METHODOLOGY

This chapter will describe the research method and its classifications. To this end, the characterization of the research will be pointed out, in order to identify its classifications, approach, means and objectives. In the universe, sampling and sample, explaining the number of individuals surveyed.

It also described the data collection instrument used and its analysis perspectives. This gave credibility to the information in the study.

2.1 CHARACTERIZATION OF THE RESEARCH

Research is a rational and systematic procedure that aims to provide answers to the problems posed. From this perspective, this research can be classified as applied and its nature is characterized as field research. Its main objective is to observe the facts as they occur in the natural environment, without isolating or controlling the variables (GIL, 2010). In this way, knowledge will be generated for the application of practices that will solve specific problems such as the adaptation of the use of information technologies in the generations studied.

As for the approach to the research problem, this study is classified as quantitative, as it measures and describes the information collected using numbers, graphs and tables (MARCONI; LAKATOS, 2010). In this way we can measure the habits and attitudes of the generations.

In addition, the research is classified as exploratory in nature, as there are few areas of scientific study that have set out to investigate and analyze the behavior of the *Baby Boomer*, X, Y and Z generations in relation to the advancement of technology. In addition, this study shows how people live with the innovations used in their lives and work in order to provide greater familiarity with the problem.

As for the means, it is classified as bibliographical through a review of the literature published in scientific studies, such as: books, newspapers, magazines, websites, among other existing sources. There was also a need to use international sources such as reliable *journals with a* strong impact in the field. This ensures that the results of this work are more serious and robust.

In terms of objectives, it can be defined as descriptive, as it studies and describes the characteristics of a given population. In this case, it will be the generations (GIL, 2010). In addition, the study will describe the analysis of the data collected by means of a questionnaire applied to a pre-defined sample, in order to show the adaptations of the human being in the use of technologies.

The scientific method used is inductive, as it is based on the knowledge of generational experience, observations of the behavior of concrete reality and which are analyzed based on particular findings. According to Marconi and Lakatos (2010, p. 68) the inductive method is: "a mental process through which, starting from particular data, sufficiently verified, a general or universal truth is inferred, not contained in the parts examined".

2.2 UNIVERSE, SAMPLING AND SAMPLE

The research universe consists of individuals born after the Second World War, as they witnessed the arrival of the computer in the 1940s, and their descendants with the evolution of information technologies born up to 2002. A research universe or population can be defined as a set of animate or inanimate beings that have at least one characteristic in common (MARCONI; LAKATOS, 2010). For the purposes of the methodological procedures required for this scientific investigation, the delimitation of 200 (two hundred) individuals was considered to represent the sample.

The sampling chosen was probabilistic, as it is subjected to statistical treatment that allows sampling errors to be compensated for and the respondents chosen to be randomly analyzed, i.e. each member of the population has the same probability of being chosen (MARCONI; LAKATOS, 2010). The criterion for using this type of sampling is to obtain the most effective and accurate information, both in terms of application and responses.

The sample is a small selection or a subset of the chosen universe or objective extracted from the entire collection (KOKOSKA, 2013). In this study, the defined sample was two hundred (200) respondents. To this end, it was previously stipulated that the application of the survey instrument would be divided into equal quantities by generation. Therefore, each of the four generations received fifty (50) respondents.

2.3 DATA COLLECTION INSTRUMENT

Two scales validated and adapted for this research were used for the data collection instrument. One of the scales used was the Resistance to Organizational Change (RAM) scale developed by Bartolotti in 2010, which includes indicators of acceptance, indifference and resistance, adapted from the author's questionnaire. Firstly, we can understand that resistance to change refers to the behaviour of an individual or a group of people, and their reaction to a certain new situation, whether inside or outside an organization. Therefore, the resistance to change scale used in this study was used to measure the forces of resistance to accepting or being indifferent to the implementation of a new information system (BARTOLOTTI, 2010).

This scale was used to measure the level of resistance of the generations to the use of technology. Table 1 shows the constructs and statements used in the questionnaire applied in this study:

TABLE 1: Likert scale on resistance to change.

Construct	Statements about resistance to technology	DT	DP	I	CP	CT
Acceptance	1. I am able to adapt to technological changes when they occur.	1	2	3	4	5
	2. I actively cooperate to achieve the change when it happens.	1	2	3	4	5
	3. I am more likely to accept a technological change when I receive information about it.	1	2	3	4	5
	4. I believe that technological changes are a way of acquiring more practicality in my daily life.	1	2	3	4	5
Indifference	5. When technological changes happen, I try to do only what is necessary.	1	2	3	4	5
	6. I prefer to remain indifferent to technological changes.	1	2	3	4	5
	7. I prefer to do the same things in my daily life, rather than try different things.	1	2	3	4	5
	8. If technological changes happen, I don't feel compromised.	1	2	3	4	5
Resistance	9. I feel that technological change is a threat.	1	2	3	4	5
	10. I feel that technological changes in my routine affect my daily life.	1	2	3	4	5
	11. If technological change means doing something I don't like, I do it slowly.	1	2	3	4	5
	12. I'm not interested in carrying out activities that will result in technological changes.	1	2	3	4	5

SOURCE: Adapted from Bortolotti (2010).

The constructs used in the scale sought to verify the behaviors of resistance to change. According to the author Bortolotti (2010), there are three (3) levels of behaviors and each level verifies indicators, which were used in this study:

- **Acceptance:** Checks whether people demonstrate acceptance through behavior, have feelings in favor of change and have conceptions, beliefs in favor of change;
- **Indifference:** Checks if they are apathetic towards change, they only do what is necessary, what is ordered, but without complaining, they are not against it, but they are

not in favor of change either, which has a great negative effect on change;

- **Resistance:** Verifies the support of actions against change and gets involved in open demonstrations looking for ways to stop it.

The second scale used was the System Usability Scale (SUS) created by John Brooke in 1986 at the *Digital Equipment Corporation* laboratory in the UK. The SUS scale evaluates the use of a product, how people feel about using it, checking whether it is easy to learn, efficient and satisfying to use (SORDI; MEIRELES 2010).

According to Brooke (1986), the SUS is a 10-item questionnaire designed to measure:

- **Efficiency:** Checks the ability of users to complete tasks using the systems and the quality of the output of these tasks;
- **Efficiency:** Checks the level of resources consumed in carrying out tasks;
- **Satisfaction:** Checks users' subjective reactions to using the system.

The questionnaire used 6 items from the SUS scale and adapted them to measure the use of technologies, with indicators used such as ease of learning and satisfaction. Chart 2 shows the constructs and statements used in the questionnaire:

TABLE 2: Likert scale on technology usability.

Construct	Statements about usability	DT	DP	I	CP	CT
Ease of learning	1. I found the technology easy to use.	1	2	3	4	5
	2. I believe I would need instructions (support) for using the technology.	1	2	3	4	5
	3. I would have thought that most people would learn to use technology quickly.	1	2	3	4	5
	4. I had to learn a lot of things before I could continue to use the technologies.	1	2	3	4	5
Satisfaction	5. I would like to use technology frequently.	1	2	3	4	5
	6. I felt very confident using the technologies.	1	2	3	4	5

SOURCE: Adapted from Brooke (1986).

Both scales comprise five (5) response categories, according to the Likert Scale (totally disagree, partially disagree, neither agree nor disagree, partially agree and totally agree). This scale can measure the level of the interviewees' behavior and directly assist in the construction of the data collection instrument.

In addition, questions on personal characteristics, socio-economic attributes and statements expressing the attitudes of the generations towards the technological reality were also used, with related closed and open questions. The aim was to assess how the

Baby Boomer, X, Y and Z generations adapt to the evolution of information technology. It is worth noting that the identity of the respondents has been preserved.

The data collection instrument was digitized on the Google Docs platform. The platform facilitated the application of the structured questionnaire applied in this study. In addition, the survey link was published on social media, specifically *WhatsApp, Facebook, LinkedIn* and email, with an eye on the target audience for this study.

2.4 DATA ANALYSIS PERSPECTIVE

Considering that the aim of this study is to analyze the evolution of the use of information systems by generation, in terms of their development and the contribution of information technology. The observations and the application of the questionnaires were carried out simultaneously and, in order to minimize the effects of the observations, notes were taken.

Quantitative data was collected electronically and analyzed using descriptive statistics (frequency, mean and percentage), with the results presented in tables to show the results more effectively. The quantitative results were correlated with the observations made. Microsoft Excel 2007 was used to tabulate the data, generating the graphs presented in the analysis with the responses of those surveyed.

3 INFORMATION TECHNOLOGY AND ITS EVOLUTION

The importance of information is due to the control it exerts over activities, communication, making the right decisions, whether related to the business or personal sphere. Provided with stored data and searched for when necessary, information is the transformation of data into something useful.

According to Penedo (2015), we can define Information Technology (IT) as a set of activities and solutions provided by computer resources aimed at enabling the obtaining, storage, access, management and use of information or even as an area of knowledge responsible for creating, administering and maintaining information management through devices and equipment for accessing, operating and storing data, so that it generates information for decision-making.

In light of this definition, Sousa (2010) says that since information technology was introduced to the world around the end of the 1950s, everything has changed and this is indisputable. Companies, for example, have started to use these technologies to innovate in the market, to produce at lower cost and in a more efficient way, thereby changing the world with impressive innovations. Information technology is causing the boundaries between work and home to become increasingly blurred.

Chart 3 below was created by Pacheco and Tait (2000), based on two authors, Reinhard (1996) and Brito (1997):

TABLE 3: History of the use of IT.

Decade	Features	
	Reinhard (1996)	**Brito (1997)**
1960	• Companies start using IT; • Few technological options (software and equipment); • Laborious application building processes with little tool support; • Need for methodologies to meet demand quickly; • Automation of manual routines; • Shortage of technical labor; • Development with character artisanal.	
1970	• Increased impact of systems on companies; • Analysts now consider: concepts of organizational development, decision-making,	• IT as a strategic organizational resource; • The era of data processing;

	adoption of innovations, learning, human-computer interface, relationship between IT professionals and users; • Encouraging the construction of decision support systems.	• IT resources as a business support tool.
1980	• Changes in companies' external environment; • Outsourcing, inter-organizational systems; • Systems architecture; • Development of systems taking into account economic, legal, political and cultural aspects.	- Business execution is increasingly dependent on the application of IT.
1990	• IT as the center of business strategy; • Knowledge as a source of value generation.	• IT becomes more strategic; • IT drives business transformation.

Source: Adapted from Pacheco and Tait (2000).

Table 1 shows the use of Information Technology from the 1960s to the 1990s, from two different points of view. Reinhard (1996) looks at the use of IT corresponding to the situation in companies, while Brito (1997) draws a parallel between IT and its relationship with business.

According to Meirelles (1994, apud SOUSA, 2010), an unprecedented evolution has caused a gigantic and dynamic increase in the applicability and performance/cost of information technologies, which has increased strategic opportunities for many organizations.

The author Veiga (2007) makes a historical analysis of information systems, showing their evolution over time, as shown in (Figure 1):

FIGURE 1: Historical analysis of information systems.

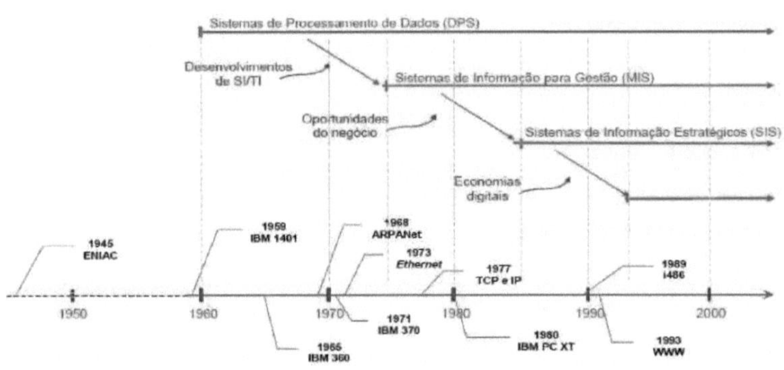

Source: Veiga (2007).

The author Veiga (2007) shows that in the 1960s the development of information systems began with the Data Processing System (DPS), in the 1970s business opportunities began and the Management Information System (MIS) was created and around the 1990s the Strategic Information System (SIS) arrived.

3.1 THE EVOLUTION OF TECHNOLOGY IN PEOPLE'S LIVES

The first computer appeared in the Second World War, created by German engineer Konrad Zuse in the 1930s and called Z1, with the functionality of formulating strategies, after the war it was created to serve the industrial market used in large companies, as they were huge and had very high financial resources (MESSINA, 2017). Figure 2 shows the computer developed for industrial companies:

FIGURE 2: Manchester Baby computer from 1948.

Source: Messina (2017).

The personal computer was created in the 1970s by Apple Computer, by businessmen Steven Wozniak, Steven Jobs and Ron Wayne, who created the Lisa project, which was the first PC to have a mouse and a graphical interface, in 1983 (PEREIRA; CHIRIU; PEDROSA; LACERDA; FRANCO; LUIZ E SILVEIRA, 2006). Since then, computers have evolved, reduced in size and their resources have become more advanced and are now part of people's daily lives, being used for work, entertainment, academic research and leisure.

The emergence of television came about through various discoveries by scientists in the 19th century, the first being the Swede Jakob Berzellus with the discovery of photosensitivity. A few years later, the German Paul Nipkow, in 1884, was the founder of the TV technique, and so other scientists developed the photoelectric cell. The scientist Constantin Perskyi, at a congress in Paris in 1900, demonstrated a piece of equipment that worked on the basis of selenium photoconductors that transmitted images over a distance, which gave rise to the word Television, and over the years this technology was perfected and the first program to be broadcast took place in England in 1930 and color

TV in Brazil in 1962 (ABREU; SILVA, 2011). Nowadays, with digital technology, we can even interact with the TV, access the internet and browse websites as if it were a computer.

Radio emerged after the development of telegraphy and radiocommunication. Between 1893 and 1894, the Brazilian scientist Landell de Moura was the pioneer in transmitting the human voice wirelessly in Brazil; there are other scientists around the world who are also considered to be inventors of radio (FERREIRA, 2013). According to Martins (1999, p.47) "Advertising reaches the radio, between chapters of soap operas, the news or in the course of programs, gracing the ears". Today *we* can tune in to the radio on various stations, on mobile devices, and with the use of the internet *we* have *webradio,* the radio is widely used for commercial advertisements, to be disseminated in the streets of cities and people from previous generations still usually listen to the news on the radio.

The telephone arose from the need to transmit information as far back as antiquity. In 1837, a monk from Citeaux Abbey created a system that transmitted words using tubes. In France, Samuel Morse invented the electric telegraph, which helped in the future invention of the telephone. After the world was already enjoying this technology, the Scotsman Alexander Graham Bell created the telephone in 1876, which arrived in Brazil in 1877 by King Dom Pedro II (NETO, 2017). Nowadays, landlines are more commonly used in companies, and with the arrival of cell phone technology, they have been somewhat neglected in homes, so we can see that communications technology has evolved greatly.

The cell phone (cell phone) can be defined as a communication technology device, which has a set of functions such as a telephone, computer, digital camera for photos and videos, word processor, GPS, among others; Mobile because it works over digital wireless networks, can use several networks and among other features (LEMOS, 2007). Nowadays, the cell phone, or what is now known as a *Smartphone* due to its technology, is present in the daily lives of people who use it for various functions, especially generation X and Y.

The origins of the Internet began around 1964, during the Cold War between the United States and the Soviet Union, the Pentagon asked Rand

Corporation to create a communication network that would be nuclear bomb-proof, scientists soon began to study the proposal, but it was around 1973 with Vinton Cerf, who created the Internet with the conditions proposed by the Pentagon (MATTOS, 2010). Since then it has evolved with various discoveries of applications such as e-mail and others. Chart 4 shows how the Internet was the breakthrough technology that reached a higher rate of users than other technologies:

TABLE 4: The meteoric success of the Internet.

How long it took to reach 50 million users	
Cars	55 years old
Electricity	46 years old
Telephone	35 years old
Microwave oven	30 years
Television	26 years old
Radio	22 years old
Microcomputer	16 years old
Cell phone	13 years
Internet	4 years

SOURCE: Veja, Sao Paulo, July 29, 1998, p.36.

The culture of the Internet is called *Cyberculture, and* author Levy (1999, p.17) defines *Cyberculture* as: a "set of techniques (material and intellectual), practices, attitudes, ways of thinking and values that develop along with the growth of cyberspace". This term describes the cultures and relationships that people have virtually, exchanging information and files.

3.2 THE POPULARIZATION AND IMPACT OF INTERNET USE ON SOCIETY

The speed with which the Internet has entered people's daily lives has had a number of impacts on cultural behavior. According to the author Mattos (2010), there are some applications that have arisen along with the internet that society uses:

- International collaboration (such as projects developed in several countries at the same time by the automotive industry);
- E-mail;
- Distance learning and training (virtual university);
- Software distribution, both paid and free (freeware, GNU);
- MP3 music;
- Scientific research through *newsgroups* (discussion boards);
- Browsing libraries and downloading books (like Project Gutenberg - http://gutenberg.net);
- Customer services (SAC);

- E-commerce (buying and selling over the internet);
- Political and citizens' movements (usually via e-mail);
- Daily news from newspapers such as Edupage - www.educause.edu/pub;
- Radio and TV;
- Chat;
- Leisure, games and entertainment.

These are some of the changes that have taken place, because the Internet service is evolving more and more and people are having to adapt to new resources. As you can see, the Internet has changed the way businesses operate and has transformed society. A survey carried out by e-commerce.org on the percentage of the population using the Internet in some countries in 2003 is shown in Table 5:

TABLE 5: Internet penetration in some countries.

Parents	**% of population using the internet (2003)**
Sweden	68,5
United States	60,1
Norway	59,1
Australia	53,8
South Korea	53,0
Switzerland	52,7
England	50,8
Germany	50,2
Japan	44,1
Portugal	43,7
Taiwan	28,7
France	28,4
Brazil	8,0
Argentina	5,3
China	3,5
Mexico	3,4
Russia	1,2
india	0,7
Irâ	0,6
Syria	0,3
Uganda	0,2

Nigeria	0,1
Chad and Ethiopia	0,0

SOURCE: e-commerce.org.br (2017).

Statistical data from the International Telecommunication Union (ITU), a body linked to the United Nations (UN), in 2015 shows that 3.2 billion people in the world were connected to the internet. According to the survey, in 2000, only 6.5% of the world's population were internet users. In the year of the survey, this figure was 43% and there were 4 billion people disconnected from the internet (ITU, 2015). The data shows the great advance of the internet and that society is becoming increasingly connected, especially in developed countries.

The internet has facilitated and changed some behaviors in people's daily lives. According to Mattos (2010, p.70): "the internet is changing the way people relate to each other, the way companies work, the way we see the globalized world". Nowadays, we can do various activities that it would take hours to be able to carry out. With the use of the internet, there are online systems that make our lives easier, let's take a look at some examples:

- **Bank branch:** With this modernization, society has benefited greatly from technology, especially the most recent generation. In bank branches, customers don't have to go to the bank to make a simple transaction, such as making a transfer, making bank payments, checking their account and so on;

- **Online shopping:** The ease of being able to store from home or wherever you are, via your computer, *tablet* or *smartphone, has* saved people time in their daily lives when they couldn't go to a physical store to buy their products;

- **Distance learning:** Nowadays, colleges or educational institutions offer people higher education or qualification courses on online platforms, as well as offering research in online libraries, so that society can update and improve professionally at a low cost;

- **Social networks:** A tool for communication and interaction with a society that is connected to the globalized world, social networks are present in people's professionalism and also in their entertainment;

- Newspapers: The high-speed news that allows us to acquire information from around the world, without having to buy a traditional newspaper, saving time, money and preserving the environment with the manufacture of paper.

4 THE STUDY OF GENERATIONS

A generation can be identified as a group of people who experienced the same birth years, the social events that marked their behavior and were significant for their development in society (KUPPERSCHMIDIT, 2000). According to some authors, generations can be characterized as a social phenomenon because they are products of historical events that have had a major influence on values, visions and principles (CHIUZI; PEIXOTO; FUSARI, 2011).

Therefore, this topic will study the profiles of the *Baby Boomers*, X, Y and Z Generations, so that you can understand how a generation differentiates its behaviors, so you need to understand how each of them forms a group of individuals with values, beliefs, customs and priorities.

4.1 *BABY BOOMERS* GENERATION

The word *Baby Boomers* appeared in the United States after a demographic explosion with an increase in the birth rate. According to some studies, the *Baby Boomer* generation emerged after the Second World War with people born between 1940 and 1960 (SERRANO, 2010). The *Boomers* experienced a world of economic prosperity, as this was the post-war period when major technological advances were made, such as the creation of computers (ROBBINS; JUDGE; SOBRAL, 2010).

Between the 1960s and 1980s, these people began to enter the job market and had to adapt to the development of technology, which made new opportunities and careers possible with the arrival of the computer (BALASSIANO, 2009). For this author, there were still those who were more resistant to this technological novelty, which was a major challenge to overcome when using the computer.

This generation has certain characteristics, such as the search for economic opportunities in various areas of social work. They dedicated their academic efforts to careers that promised ease in the search for positions within a business organization, always valuing *status and* professional advancement within the company (COIMBRA; SCHIKMANN, 2001).

The author Castro (2011) published some of the main characteristics that can be determined in the *Baby Boomer* Generation, which are:

1. It has a more consolidated income;
2. It has a more stable standard of living;

3. Little influence from the brand at the time of purchase;
4. It values experience;
5. They have a greater preference for high-quality products;
6. He prefers quality to quantity;
7. Not easily influenced by other people;
8. He doesn't see price as an obstacle to pursuing a desire;
9. He is firm and mature in his decisions.

Some important events occurred during the *Boomers*' period, as they emerged at the time of globalization, which was marked by facts such as: man going to the moon, capitalism and consumerism, rock and roll, the Hippie movement, political and social protest, peace movements, the Vietnam War, libertarian ideology and feminism, promoting labor conquests, among many other movements that changed society (CASTRO, 2011). In addition, the *Baby Boomers* conquered various social causes and were great agents of transformation, including: the breaking down of political barriers regarding the role of women; young people leaving home to live alone; raising awareness of peace, love and free sex. Thus, this generation was marked by various demands that catalyzed a series of changes, which are still going on today (OLIVEIRA, 2013).

These major events have influenced this generation in their behavior at work, they have acquired a great sense of justice, transience or authoritarianism and loyalty to their company, their professional career and relationships in the workplace (WEINGARTEN, 2011). They are dedicated to their position in the company and achieving their goals, and have difficulties balancing their professional and personal lives.

4.2 GENERATION X

Generation X emerged in the 1960s to 1980s and are considered to be the children of the *Baby Boomer* Generation. They had the opposite thinking to their predecessors, who fought for freedom of expression; these young people had the characteristic of seeking to value the opposite sex and demanding their rights (OLIVEIRA, 2013). These people are conservative, self-confident, believe in their own self-esteem, are very creative, always meet their objectives without respecting deadlines and have an aversion to supervision (LOMBARDIA, 2008).

This generation was prepared to live with the absence of their parents, as they dedicated themselves to their professional lives, which led to many divorces in the *Baby Boomer* generation. As a result, Generation X grew up with family behavior unlike that of their parents, dedicating themselves more to personal than professional relationships, generating flexibility between these relationships (ROBBINS; JUDGE AND SOBRAL, 2010). For these authors (2010, p.141) Generation X had the following behaviors: "balanced lifestyle, teamwork, rejection of norms, loyalty to relationships".

Some of the characteristics of this generation were birth control, families chose to decide on a smaller family formation, they were more controlling in terms of consumerism, always looking to save money, they weren't as dedicated to organizations as the *Baby Boomers* and they were more independent (GLASS, 2007). They were quick to change, so the tradition of the previous generation could lose its customs passed on to the new generation and also in professional matters they value recognition for merit and not for length of service, they like to carry out their activities individually and they like more informal arrangements at work (SULLIVAN et al., 2009).

Generation X experienced major events in the information system, such as the launch of color TV, the personal computer and the Internet. They entered the job market between the 1980s and 2000s and were already using computers and the internet as a working tool (VESCOVI, 2012). In addition, this generation preferred organizations that guaranteed the development of skills, productivity and balance in their personal and professional lives (CENNAMO; GARNER, 2008). Thus, it can be said that they were individualists who did not depend on the opinions of others.

4.3 GENERATION Y

Generation Y, considered the daughter of the previous generation (Generation X) and granddaughter of the *Baby Boomers*, also referred to by some authors as *Millenials*, was born between 1980 and 2000, in a world scenario of transformation due to the process of globalization. For people born in this generation, technology and communication have always been part of their lives and experiences, unlike their predecessors (where there was a process of adaptation and acceptance). In the Information Age, this context of systems and technology development was already a reality.

Serrano (2010, *apud* Naconeczny; Santos and Baggio) points out that Generation Y is a relatively new generation, with no clear conceptualization of its characteristics yet, except for the fact that they were born into a world that was becoming one big global network; where the Internet, emails, social networks, digital resources, influenced the generation to use the opportunity to make thousands of friends around the world, without even leaving their computers.

According to Vescovi (2012), some of the characteristics of Generation Y are defined: "Speed and ease of adapting to change, agility in solving problems and promptness in presenting answers; the ability to analyze and react in risky situations were some of the behaviors identified in this generation." These individuals are anxious, i.e. impatient and want to acquire products or benefits as quickly as possible, disrespecting hierarchies in organizations (ROBBINS; JUDGE AND SOBRAL, 2010).

Some of the characteristics of Generation Y individuals are listed in Chart 6:

TABLE 6: Characteristics of Generation Y.

characteristics of generation y	
1	The multitasking generation.
2	Seekers of innovation and technology.
3	They share their lives, data, photos and habits through technology.
4	Generation always present in social networks.
5	They prefer typing to writing, e-mail to letters.
6	They choose computers over books.
7	They seek information as easily and immediately as possible.
8	Always connected, otherwise they experience withdrawal.

Source: Saviel (2009), adapted by the author.

4.4 GENERATION Z

Generation Z is the generation born from the 1990s onwards and is known for being born during the advent of the internet and the high technology brought about by post-modernity. The world of this generation is technological and virtual, as it is full of information, since everything that happens is reported in real time via the internet and high-quality technology, often making this immense volume of information obsolete in a short space of time (CIRIACO, 2009).

The generation is known as Z because of the verb zapear, which is used to show the act of changing the channel on the television using a remote control (CEREITA E FROEMMING, 2011). This characterizes their ability to reconcile doing several activities at the same time and being more adaptable than Generation X.

This Generation Z doesn't care about hierarchy or qualifications, and they don't think that life is just about a career. In fact, life only makes sense if they do what they love. Their main character is entrepreneurial and self-taught, as almost 80% of them produce and share content. This encouraging outlook has already brought about profound changes in the job market, business models and management (ANAUTE, 2017).

It can be said that the difference between Generation Z and other generations is the ease with which they use technology and adapt to technological changes, however difficult they may be. However, this generation is also known as Silent, as it is defined as one that tends towards self-centeredness, as it is concerned only with itself (INFOMONEY, 2010).

5 GENERATIONS TODAY

This topic shows the characteristics of the generations today, how they are adapting to new technological developments, their lifestyles, age groups, socio-demographic information and the current study of generations in the job market and in business organizations. Before detailing this information, a summary of the main characteristics of the *Baby Boomer*, X, Y and Z generations is presented in (Chart 7):

TABLE 7: Main characteristics of the generations.

GENERATIONS	Baby Boomers	Generation X	Generation Y	Generation Z
Born	1945 a 1964	1965 a 1980	1980 a 2000	From 2000
Job market	1965 a 1985	1985 a 2000	From 2000	-
Social and economic context	Urbanization; Rights movements (civil, feminist and dictatorship); Entry of new technologies.	Globalization; His parents' excessive working hours; Launch of cable TV; Using computers as a working tool; Socialization of the Internet.	Too much information; Technologies had already been absorbed and became part of people's daily lives; Social networks; Electronic games.	The digital world in people's lives.
Features	Rejection of authoritarianism; Personal fulfillment; Material success; Ends justify the means (Pragmatics); Ambition.	They are self-sufficient (absence of parents). Independent; Balance between personal life and work; They value flexibility, satisfaction in the workplace and knowledge	Prosper quickly; Constant opportunities and learning; Quick and easy to adapt to change; Agility in solving problems and submitting	They are dynamic and innovative; Concerned about the environment; Easy to do several tasks at once; They are critical and change their minds several times;
		technical.	ability to analyze and react to risk situations.	They will be more demanding professionals.

SOURCE: Vescovi (2012), adapted by the author.

According to a study carried out by TNS Brasil (2016) with around 60,000 respondents from 50 countries, a survey was carried out on the habits of the *Baby Boomer*, X and Y generations in relation to the means of information systems used. The difference in how many hours the generations are connected and the means used to acquire information are shown in Table 8:

TABLE 8: Generation gap: data from the TNS Brazil study.

Device/activity	Baby Boomers	Generation X	Generation Y
Time spent daily (hours)			
Mobile	1.5	2.4	3.2
TV	2.3	1.3	1.3
Radio	0.5	0.2	0.2
Newspapers and magazines	0.3	0.1	0.1
Watching videos and TV online	1.2	1.6	1.9
Social media	1.2	1.6	2.4

SOURCE: TNS Brasil (2016).

This TNS survey only confirms the reality of Generation Y, who are connected 3.2 hours a day on their cell phones and more on social media. Compared to previous generations, the younger they are, the more they are connected to the virtual world.

5.1 *BABY BOOMERS* IN THE AGE OF TECHNOLOGY

The current era is known as the 'Age of Technology', where information systems are constantly evolving, such as digital image telecommunication systems, video calls on *smartphones* and online banking payments. The *Baby Boomer* generation currently has different habits from the past, but they are also different from Generation X, in aspects such as concern about food, aesthetics and physical activities (BARBIERI, 2015).

Today's *Baby Boomers* are between the ages of 57 and 77 and they prioritize quality of life, exploiting past experiences.

as a reference for the future, they are not influenced by the current generation, do not see price as an obstacle but rather the desire to buy, are firm and mature in their decisions, have a more stable income, have a secure standard of living, suffer little influence from the brand at the time of purchase and are impatient with products and services that do not deliver what is promised (BARBIERI, 2015).

According to some authors, this generation currently occupies important positions in their professional lives, due to their high level of control and command in their leadership style, they have reached high levels of hierarchy in organizations and command large work teams (WEINGARTEN, 2011; SOUZA, 2010).

We can divide them into three current profiles: Those in the 50 to 60 age group are considered young in spirit, with mature bodies and thoughts. They like to travel, make friends and make the most of lost time with the security that comes with age. Some prefer to experience the new and understand that there is a lot to explore and others value good service and agility, are more demanding and have more criteria for choosing what to do, where and what to spend (VECCHI, 2015). They are people who respect family traditions and value hierarchical positions within organizations, good interpersonal relationships at work, mutual exchanges and values of reciprocity (CHIUZI, 2012).

5.2 THE COEXISTENCE OF GENERATION X IN THE TECHNOLOGICAL WORLD

Generation X is a group that has followed technological advances, changes in behavior, different political moments, all of which and many other factors influence this change (HELABS, 2017). These movements have had to undergo frequent adaptations so as not to lose their place to the more recent and up-to-date generations and still maintain their essence and traditions.

This generation today are more cautious people, they limit themselves to the age of technology, they only use essential communications for work such as email, some social networks, search engines like Google, they still prefer to read newspapers, watch advertisements on the radio and television (HELABS, 2017).

They are currently aged between 37 and 57. In the job market, they like challenges, a variety of roles, new opportunities, they want to work with freedom and flexibility and always need *feedback to* improve their weaknesses. They are skeptical and like a more informal work environment with a less strict hierarchy, i.e. a job with less bureaucratic activities (COMAZZETTO; PERRONE; VASCONCELLOS E GONÇALVES, 2016).

5.3 GENERATION Y AND THEIR CURRENT HABITS

Currently they are in the 28 to 37 age group, individuals who were born during the evolution of technology, they share the same work environment as previous generations, generating conflicts and increasing the turnover rate in companies, at the same time their contributions are essential in activities that use technologies, since they have more skills in handling technological tools (FLINK; FERREIRA; HONORATO; ARAUJO E PROENÇA, 2012).

These young people have a difference compared to Generation X, they have a faster learning capacity, less prejudice towards diversity, a topic that is currently being discussed, they have a more globalized mindset and are always constantly changing and innovating (OLIVEIRA, 2015). They are impatient, stressed, anxious and ambitious, as they have been influenced by the internet.

According to a survey by Business Insider (2016), it lists 7 negative habits of Generation Y:

1. Few hours of sleep: Change in sleep schedule, with different times and irregular hours. In addition, they often use laptops or *smartphones* for a long time before going to bed, which affects the quality of their sleep;

2. Lack of regular eating: Lack of routine and frequently changing schedules can affect the regularity of eating. Going too long without eating or drinking can lead to low sugar levels, causing anxiety, dizziness, shakiness and even loss of attention;

3. Coffee: Although it has some positive effects, excessive consumption can make you more nervous and irritable. It can also cause dehydration, which leads to anxiety;

4. Spending a lot of time sitting down: According to the study, the risk of anxiety increases in people who lead a more sedentary lifestyle. But the most interesting point was that the more time you spend sitting down, the more likely you are to suffer from anxiety;

5. Mobile: According to a study carried out by researchers from American and Singapore Universities, Facebook can cause envy and even depression in people, as it can lead them to compare their lives with those of others;

6. No disconnect: The technology so present in the lives of *millennials* makes it harder to "disconnect" from work, even after the working day. This invasion of professional life into personal life causes irritation and anxiety. "*Millennials* don't believe that productivity should be measured by the number of hours worked in the office, but by the work done. They see work as a 'thing' and not a 'place'," explains the BDA survey.

7. Streaming services: The University of Toledo in the USA analyzed the habits of more than 400 people, 70% of whom watched TV for at least 2 hours. The result? These volunteers felt much more anxious and depressed than those who spent less time in front of the TV. The feeling of relaxation that this behavior causes is short-term.

Throughout history, this generation has shown that they seek personal success, as well as professional success, without fear of taking risks and leaving their comfort zone. They are assertive, independent of other people's opinions, innovative and concerned about socially vulnerable groups, aware of the environment; with a perception of the right to freedom of expression and democracy; as well as being more vain every day, well aware of market and fashion trends and novelties.

5.4 GENERATION Z THE TEENAGERS OF COMPUTERIZATION

Generation Z are our teenagers of today, they are in the age group of 17, young people who were born into the technological world. They live connected 24 hours a day, this generation claims that it is impossible to live without a cell phone, computer and the internet. According to the author Tapscoott (2010, p. 53):

> They want to be connected with friends and family all the time, and they use technology - from phones to social networks - to do this. So when the TV is on, they don't just sit and watch it, like their parents did. The TV is background music for them, and they listen to it while they search for information or chat with friends online or via text messages. Their cell phones are not just useful communication devices, they are a vital connection to their friends.

This generation makes even greater use of the internet for their consumption process, i.e. their purchases. According to the author Tapscoott (2010), there are some guidelines that Generation Z takes into account in the shopping process: The freedom to choose from the options on offer; the possibility of customizing/personalizing the product, making it exclusive; research into the product prior to purchase; concern about the credibility and integrity of the selling company; the possibility of making suggestions and collaborating to improve the goods and services on offer; entertainment and fun linked to the product; speed in meeting their needs and responding to any queries: the modernity of the products on offer, which adds status to the user, among their social group. This shows the great influence of information on the purchasing power of individuals who live connected to the Internet.

The personality of this generation has very strong characteristics since childhood, because they have the power to give opinions and decide what they wear, they make decisions quickly, because their minds are already used to playing video games and this requires them to have decision-making skills. They therefore have the ability to acquire information more quickly than previous generations (KULLOK, 2017).

6 Analysis and discussion of results

This section will analyze the results of the 200 respondents who are the *Baby Boomers*, X, YeZ generations, verifying the behavior of these generations in relation to their use of information systems. Analyzing their profile, their resistance to technological change and their usability of information systems. In order to achieve the aim of the research, which is to analyze the evolution of the use of information systems by the Baby Boomers, X, Y and Z generations.

6.1 PROFILE OF GENERATIONAL RESPONDENTS

The tables below show the profile presented by respondents from the *Baby Boomer*, X, Y and Z generations, with the following variables: gender, age group, level of education, marital status and monthly income. The profile of the *Baby Boomer* generation is shown in Chart 9:

TABLE 9: Profile of the Baby Boomer Generation.

BABY BOOMERS GENERATION (born between 1940 and 1959)		
GENDER	Female	46%
	Male	54%
CIVIL STATUS	Married	80%
	Divorced	4%
	Viùvo	16%
education level	No education	2%
	1st grade completed or elementary school completed	28%
	2nd grade or secondary education completed	60%
	3rd grade or higher education	10%
MONTHLY INCOME	Up to R$ 937,00	4%
	Between R$ 937.00 and R$ 1,874.00	44%
	Between R$ 1,874.00 and R$ 3.748,00	44%
	Between R$ 3,748.00 and R$ 4.685,00	4%
	Above R$ 4,685.00	4%

SOURCE: Research data (2018).

According to the data we can analyze that there was a gender balance (54%) of those surveyed were male and (46%) female, born between the 40s and 60s, currently aged between 58 and 77, considered to be "senior citizens or retirees". People who witnessed the birth of the computer, color TV and already used radio and landlines.

The majority of this generation (80%) are married and a minority are widowed and divorced. With this, it can be inferred that the majority are people with independent and stabilized lives, the author Castro (2011) mentions some profiles of this generation and two of them have a more consolidated income and a more stable standard of living. While this percentage of widowers may be related to the age group, as some of the interviewees are senior citizens and the divorcees were in relationships that somehow didn't work out.

The majority (60%) had completed high school and the other requirements were less than thirty. Considering the difficulties that existed in their time to study, the rate of secondary education was high, compared to higher education, which was low, although companies at the time didn't require higher education as they do today. This generation was characterized by the search for stability, in other words, they just wanted a stable job.

With regard to monthly income, there was an equal percentage (44%) in the R$937.00 to R$1,874.00 and R$1,874.00 to R$3,748.00 brackets. This is probably related to the level of education of this generation and their comfortable standard of living, with a small minority having an income above R$ 4,685.00, indicating that this minority are individuals with a higher level of education, which is higher education.

The results of the Generation X profile are analyzed in Table 10:

TABLE 10: Profile of Generation X.

GENERATION X (BORN BETWEEN 1960 AND 1979)		
GENDER	Female	56%
	Male	44%
CIVIL STATUS	Single	8%
	Married	76%
	Divorced	16%
EDUCATION level	1st grade completed or elementary school completed	6%
	2nd grade or secondary education completed	54%
	3rd grade or higher education	24%
	Complete postgraduate degree	16%
MONTHLY INCOME	Up to R$ 937,00	6%
	Between R$ 937.00 and R$ 1,874.00	26%

	Between R$ 1,874.00 and R$ 3.748,00	32%
	Between R$ 3,748.00 and R$ 4.685,00	8%
	Above R$ 4,685.00	28%

SOURCE: Research data (2018).

Generation X, born in the 60s and 80s, are currently aged between 38 and 57. The survey showed that (56%) of those surveyed were female and (44%) male, individuals who witnessed the emergence of the personal computer and the great innovation that was the internet. In terms of marital status, there is still a high rate of married individuals (76%). It can be seen that this generation still practices the cultural and religious customs of the previous generation, but according to the theory cited in the study they have a mentality of forming small families and dedicating themselves more to work.

In terms of level of education, there has been an increase compared to the previous generation in terms of higher education and completed postgraduate degrees. It is therefore possible to see that over the years the demand for qualifications or degrees in the professional world has become more relevant. We can see that the elementary school education rate has dropped to (6%) compared to the *Baby Boomer* generation, which was (28%).

With regard to monthly income, there was also an increase to (28%) of those with an income above R 4,685.00. Possibly due to the need for qualifications, there has been a regression in values and this generation has the characteristic of being more dedicated to their professional life. And a good proportion of the individuals have an income of between R$ 937.00 and R$ 1,874.00 and an income of between R$ 1,874.00 and R$ 3,748.00.

Data on the profile of Generation Y interviewees is shown in Table 11:

TABLE 11: Profile of Generation Y.

GENERATION Y (BORN 1980-19i)		**89)**
GENDER	Female	52%
	Male	48%
CIVIL STATUS	Single	42%
	Married	56%
	Divorced	2%
education level	2nd grade or secondary education completed	46%
	3rd grade or higher education	34%

	Complete postgraduate degree	20%
MONTHLY INCOME	Up to R$ 937.00	4%
	Between R$ 937.00 and R$ 1,874.00	30%
	Between R$ 1,874.00 and R$ 3.748,00	38%
	Between R$ 3,748.00 and R$ 4.685,00	10%
	Above R$ 4,685.00	18%

SOURCE: Research data (2018).

Generation Y, considered to be the children of Generation X, are currently aged between 28 and 37, people who were born during the evolution of technology. In relation to this generation, the survey showed a balance between female (52%) and male (48%). As for marital status, there was a high percentage of single people (48%) and a decrease in married people compared to the generation before X. This shows that this generation are more independent people who like freedom.

With each passing generation, the level of schooling increases. In this generation, there has been an increase in higher education (34%) and post-graduate education (20%). As a result, we can see that people are becoming more and more qualified in order to achieve a higher professional status and a better quality of life. In terms of income, there has been a considerable increase in people earning between R$1,874.00 and R$3,748.00, and some (18%) earn more than R$4,685.00.

Chart 12 analyzes the profile of Generation Z, who are today's young people:

TABLE 12: Generation Z profile.

GENERATION Z (BORN BETWEEN 1990 AND THE PRESENT DAY)		
GENDER	Female	74%
	Male	26%
CIVIL STATUS	Single	94%
	Married	6%
education level	1st grade or secondary education completed	6%
	2nd grade or secondary education completed	36%
	3rd grade or higher education	58%
MONTHLY INCOME	Up to R$ 937,00	14%
	Between R$ 937,00 and R$	30%

	1.874,00	
	Between R$ 1.874,00 and R$ 3.748,00	24%
	Between R$ 3,748.00 and R$ 4.685,00	14%
	Above R$ 4,685.00	18%

SOURCE: Research data (2018).

Generation Z, as the most current generation is known, is a generation of young people and teenagers who, according to the data, are currently under the age of 27, young people who were born in the age of technology, where the world lives around the internet. According to the data, the vast majority of those surveyed (74%) were female and (26%) male.

As far as marital status is concerned, most of them are single (94%). Perhaps this can be explained by their young age. Thus, the number of married people in this generation was minimal (6%). As for the level of education, as we have already seen in previous generations in terms of qualifications, which have been increasing, this generation also saw an increase of (58%) of respondents with higher education. This further emphasizes that each generation that emerges will have a higher level of education. From this perspective, it can be seen that the professional world is increasingly demanding and competition is higher, so people are looking for more qualifications.

According to the data, monthly income was fairly balanced. Thirty percent (30%) of those surveyed have a monthly income of between R$937.00 and R$1,874.00. While (24%) of the respondents have between R$ 1,874.00 and R$ 3,748.00, (14%) have between R$ 3,748.00 and R$ 4,685.00 and (18%) more than R$ 4,685.00. Possibly because they are young and at the beginning of their professional careers, others are still finishing their degrees, and some are dependent on their families, the results of the survey were diverse.

6.2 EVOLUTION OF INFORMATION SYSTEMS IN THE LIVES OF DIFFERENT GENERATIONS

The graphs and tables below show data on the *Baby Boomer*, X, Y and Z generations, how they reacted to these technologies, whether they found it difficult to handle the equipment, how often the generations use this modernity and how they enjoy it. Graph 1 shows whether the generations like the technologies or not:

GRAPH 1: Affirmation and denial of technologies.

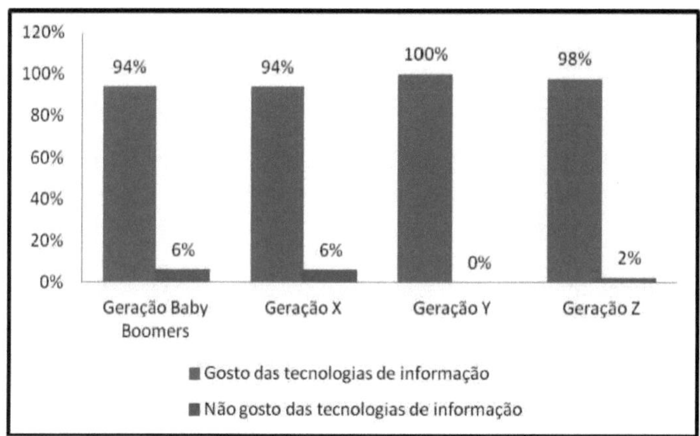

When analyzing the results of the question about whether the generations like or dislike technology, it was found that the results were balanced and showed a high index for all the generations, around (90% to 100%), who said that they like using technology, which means that technology is very present in people's lives. A small minority did not like technology, such as (6%) of the *Baby Boomer* generation, possibly in relation to people in the older age groups, who are more traditional people who are still wary of using modern technology.

With regard to the difficulties experienced by the generations in handling technological equipment, the following results are shown in Graph 2:

CHART 2: Frequency of difficulty in handling.

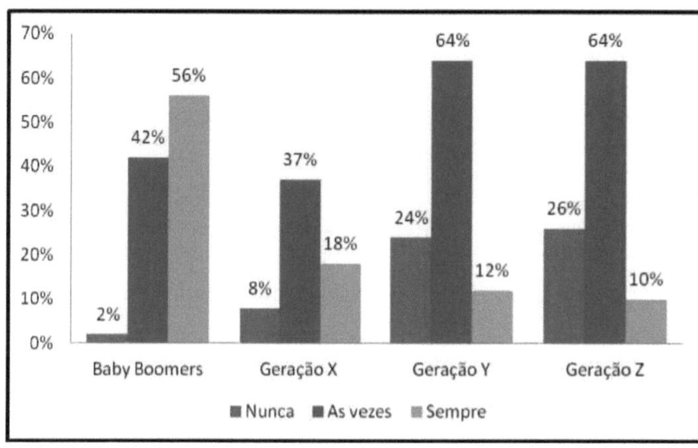

SOURCE: Research data (2018).

According to the data presented in Graph 2, it was found that the *Baby Boomer* generation obtained a higher score in the "always" requirement with 56% and the other generations with a lower score, as the *Boomers* experienced the birth of the computer and other equipment, but did not take advantage of this technology, as it was equipment that only the large industrialists of the time had. As the years went by, this ease of access grew and considerably the other generations don't feel many difficulties in their handling of technological equipment.

In the "sometimes" requirement, generations Y and Z both (64%) of those surveyed said that they sometimes find it difficult to handle equipment. Possibly because they are the youngest generations and were born in the evolution and technological era, with great ease of access to the other generations, they had a higher rate of "always" experiencing difficulties. While generations Y and Z had a higher rate of "never" with (26%) of Z and (24%) of Y, probably people more used to the information system.

There is a wide range of information technology equipment, according to the data, the most frequently used today by the generations shown in (Table 13):

TABLE 13: Information technology most used by the generations.

THE GENERATIONS				
Information technology	**Baby Boomers**	**Generation X**	**Generation Y**	**Generation Z**
Landline	0%	0%	0%	0%
Television	72%	14%	2%	0%
Radio	0%	0%	0%	0%
Computer	0%	16%	14%	8%
Mobile	20%	26%	10%	30%
Smartphone	8%	42%	74%	62%
Tablet	0%	0%	0%	0%
Others	0%	2%	0%	0%

SOURCE: Research data (2018).

It was found that landlines, radios and *tablets* are not used by all generations. Possibly because of the rush of everyday life, people no longer stay at home but at work and with the modernity of the mobile phone, people no longer buy landlines, in relation to the radio, people prefer to listen to music or news while connected to the internet and the *tablet,* because it has the same functions as a *smartphone, has* been replaced. With regard to television equipment, the *Baby Boomer* generation had the highest rate (72%). These are people who have more leisure time, retired people who like to relax by watching movies, soap operas and news programs, while the other generations had a low rate.

Computers are widely used in companies today, and Generation X and Y are the ones who use them the most. This shows that they are people who use it for their work or students doing academic research. The cell phone or mobile phone, simpler devices that don't have access to applications, showed a diversity in the results with Generation Z with (30%), (26%) the Y, (20%) the *Boomers* and (10%) the Y, which shows that this means of communication is still widely used.

The *Smartphone, a* high-tech information device with various entertainment applications and access to the Internet, was highly rated by the X, Y and Z generations. As these are people who have experienced the evolution of having a minicomputer in their hands and who need to communicate and work, this equipment was the result of this question, which the generations use most frequently. This result supports the author Tapscoolt (2010) mentioned in the theoretical framework when he says that Generation Z "want to be connected with friends and relatives all the time", via *Smartphone*.

Information technologies are used in people's daily lives in a variety of activities, and this issue was researched across the generations as shown in Graph 3:

GRAPH 3: Use of information technology.

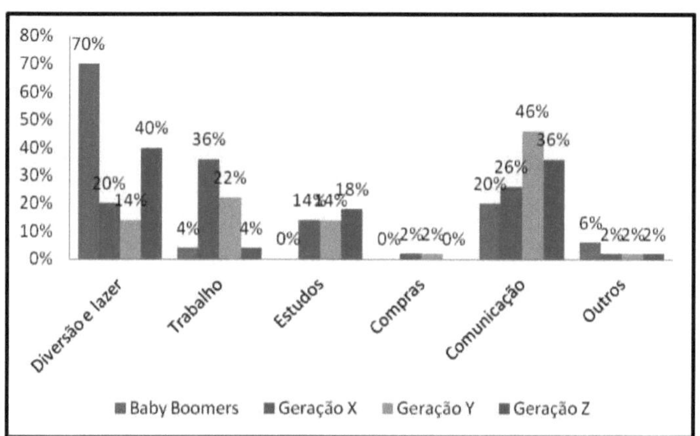

It was found that the *Baby Boomer* generation had the highest score for entertainment and leisure (70%). According to this data, it only confirms the result of the previous question in chart 13 that *Boomers* use technology for leisure. Generation Z had the second highest rate (40%), possibly because they are young and don't work yet, they also use it for fun and leisure.

In terms of work, Generation X scored highest (40%), followed by Generation Y (22%). Considerably according to their age groups, these are people who are professionally stable, and they use information technology more for work. Use for studies was low for all generations, even Generation Z, who are students.

The shopping requirement was the lowest for all generations, with only a minority (2%) of Generation X and Y respondents shopping online. As a result, it can be seen that people are still wary of shopping online, afraid of being scammed. In terms of communication, there was a balance in the results between the generations, showing that communication is the most used between them.

Nowadays most people are connected to social networks, which is why we asked how many hours a day the generations are connected and the result is shown in Graph 4:

CHART 4: Daily hours of social media use.

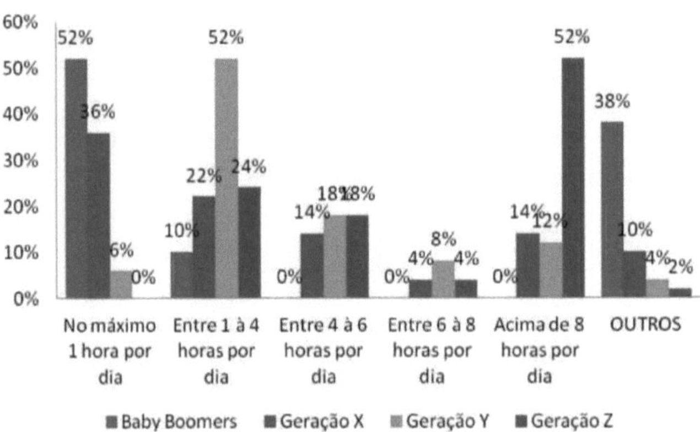

The data presented in Graph 4 above shows the use of social networks. It can be seen that all generations use social networks, which is in line with the statistical data already mentioned in the study - a survey carried out by the ITU (2015) - that 3.2 billion people in the world are connected. The majority of *Baby Boomers* (52%) are connected for a maximum of 1 hour a day and (38%) do not have social networks, as this is a generation that uses television more to obtain information, so the index was high in these two requirements.

In the requirement of between 1 and 4 hours a day, Generation Y had the highest index (52%). It can be analyzed that this generation uses it for work or personal communication, as Graph 3 shows a high index in this communication requirement. Between 4 and 6 hours a day, Generation Y and Z were tied, both with (18%) and above 8 hours a day, Generation Z with (52%) and only a small minority do not use it. As this generation was born and raised using social networks for entertainment and communication, it obtained this high index compared to the other generations which obtained a low index.

6.3 RESISTANCE TO TECHNOLOGICAL CHANGE

The graphs below show questions that verified the acceptance, indifference and resistance of the *Baby Boomer*, X, Y and Z generations to the use of information systems. Measuring through behavior whether they are interested in or opposed to technological change.

6.3.1 Level of acceptance

In relation to the ability to adapt to technological changes, the following result was presented, as shown in Graph 5:

CHART 5: Adapting to technological changes.

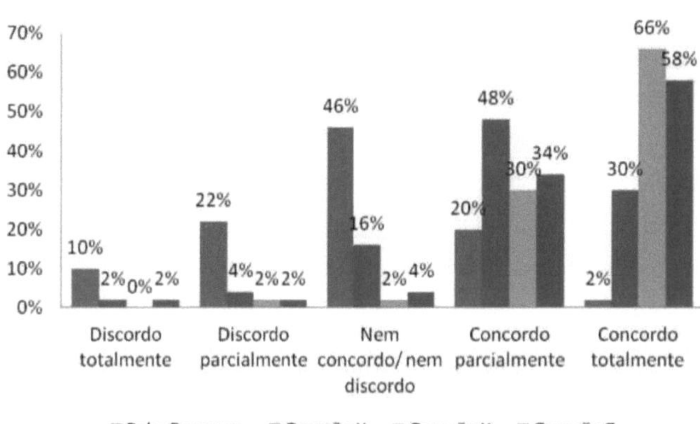

SOURCE: Research data (2018).

According to the data on the ability to adapt, there were low rates for all generations in the totally disagree and partially disagree requirements. This means that the generations feel capable when technological changes occur in their daily lives. The data showed (46%) of those surveyed from the *Baby Boomer* generation were indifferent and the other generations had high scores for the partially agree and totally agree requirements, especially the younger generations, which only confirms that the generations are not intimidated by modernity.

The respondents were asked whether they actively cooperate to make changes happen, as shown in Graph 6:

GRAPH 6: Cooperation in making changes.

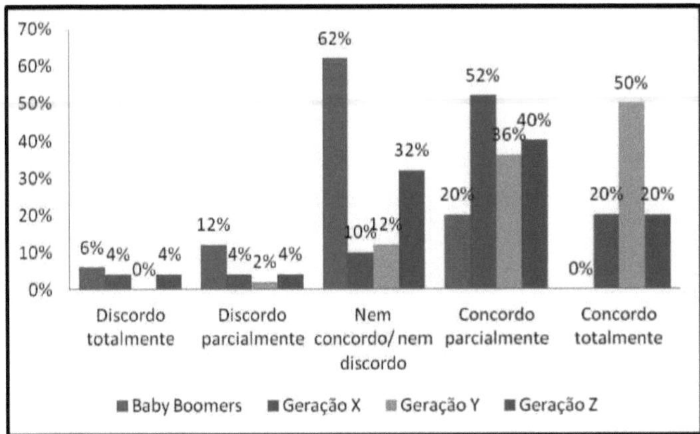

SOURCE: Research data (2018).

According to the statement, it can be seen that the majority of respondents from the *Baby Boomer* generation chose the neither agree nor disagree requirement, i.e. they prefer to remain indifferent. With regard to the partially agree requirement, there were more diverse results across the generations, which shows that the majority of those surveyed are inclined to cooperate with technological changes when they happen, as they will benefit from these changes, especially Generation Y with (50%) who totally agree.

Like Generation Y, these are people who are now adults and are in the professional world. Every year, technologies evolve and companies also acquire these modernities in order to improve the practicality of work activities. It can be said that this generation cooperates in search of practicality in their daily professional and personal lives.

In relation to the fact that generations are more likely to accept technological changes when they receive some kind of information about them, the following result is shown in Graph 7:

GRAPH 7: Willingness to accept technological change.

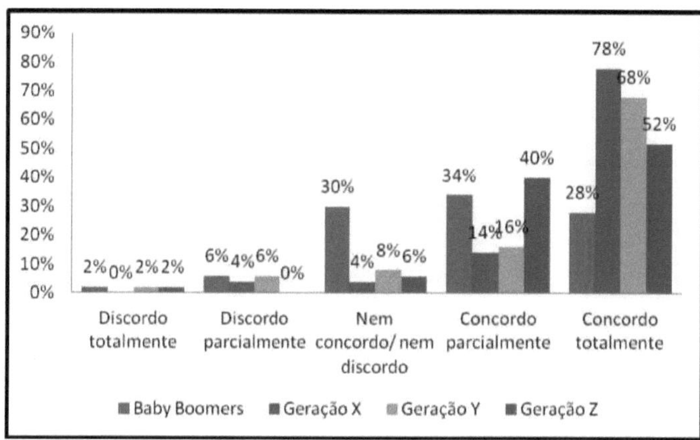

SOURCE: Research data (2018).

In view of the data, it can be seen that the majority of all generations obtained a high level of agreement with the requirement. This proves that all generations, regardless of their age, social class or education level, are more likely to accept technological changes when they receive information about them. In order to be able to use some kind of equipment, people first seek information about it.

With regard to the totally disagree requirement, the data shows a small percentage of respondents from the *Boomer*, Y and Z generations with (2%) of both. This shows that these are people who are more likely to accept technological changes and who don't need information.

Graph 8 shows that those surveyed believe that technological changes are a way of making everyday life more practical:

GRAPH 8: Technologies are a means of practicality in everyday life.

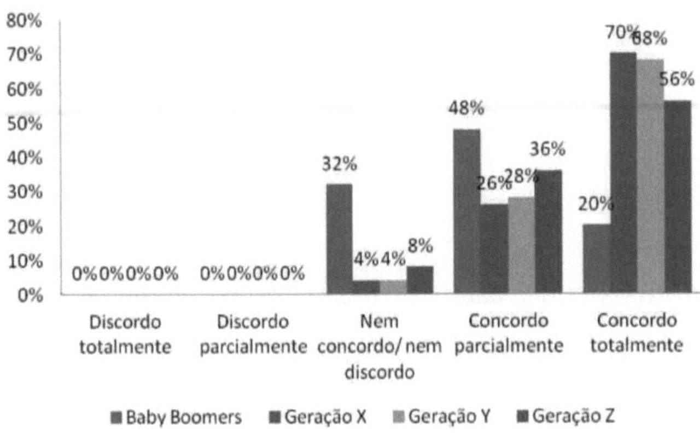

SOURCE: Research data (2018).

In view of the data in Graph 8, it was analyzed that in the totally disagree and partially disagree requirements, all the generations obtained zero results, and in the partially agree and totally agree requirements, the scores were high. This shows that all generations believe that technological changes are a way of making everyday life more practical, thus improving the execution of activities and the speed of information.

6.3.2 Level of indifference

Regarding the indifference in the behavior of the generations, when technological changes occur, they try to do only what is necessary, the following result was verified as shown in Graph 9:

GRAPH 9: Generations intimidated by technological change.

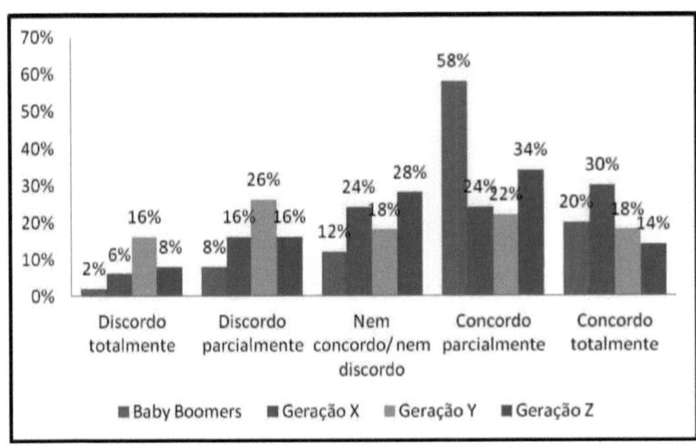

SOURCE: Research data (2018).

In view of the data, it was observed that in all the requirements there was a diversity of opinions, in the partially agree requirement where the highest index (58%) was the *Baby Boomer* Generation. It can be seen that the *Boomers*, as they are older, are more closed-minded about new things, prefer to do only what is necessary, and don't try to get involved when changes occur, while Generation Z obtained (34%), as they are young, the majority of students also prefer to be indifferent to changes.

In the case of totally disagree and partially disagree, the results were lower but similar. This minority are people who don't try to do only what is necessary, they are always looking for innovation. The results also showed a diversity of opinions who prefer to remain indifferent (neither agree nor disagree), i.e. they are people who are irrelevant to changes.

Graph 10 shows that the generations prefer to remain indifferent to technological changes:

GRAPH 10: Indifference by generation to technological change.

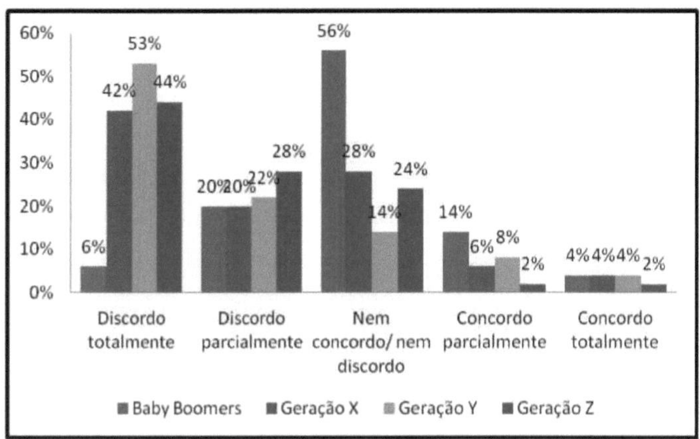

SOURCE: Research data (2018).

In view of the data, a comparison can be made with Graph 9, which showed that *Baby Boomers have* a higher rate of preferring to do only what is necessary, Graph 10 showed that *Boomers are in the* minority with (6%) of the requirement that they totally disagree with being indifferent to technological changes, while the other generations showed high rates. With this, it can be seen that the result is emphasizing the previous question that only a minority of *Boomers* have a personality that does not prefer to be indifferent to the situation, but rather in search of technological novelties, just like the other generations.

In terms of partially and totally agree, the lowest indices were found for all generations and in terms of neither agree nor disagree, the *Boomers* with 56% and the others below 30%. This shows that the majority of those surveyed do not prefer to be indifferent and another part shows a behavior that is neither for nor against technological novelties.

In relation to the fact that generations prefer to do the same things on a daily basis, rather than try different things, the following data is presented in Graph 11:

GRAPH 11: Generations don't like new experiences.

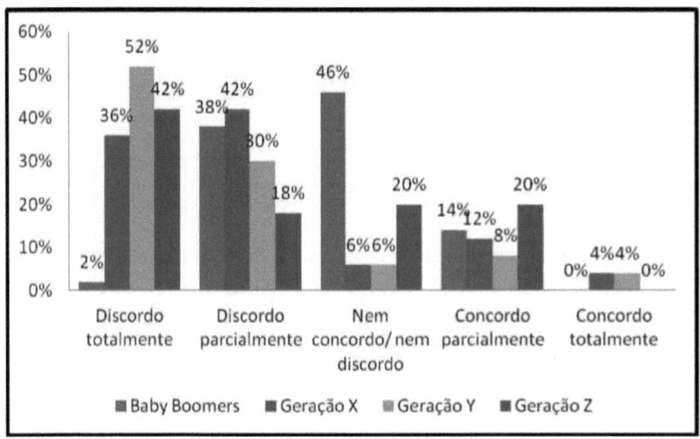

SOURCE: Research data (2018).

According to the data, the requirements totally disagree and partially disagree resulted in high rates and the requirement neither agree nor disagree with (46%) the *boomers.* This shows that the majority of those surveyed are open to new technological experiences and around half of those surveyed from the *Baby Boomer* generation prefer to remain indifferent.

With regard to when technological changes occur, I don't feel committed, the following results are shown in Graph 12:

CHART 12: The commitment of the generations to change.

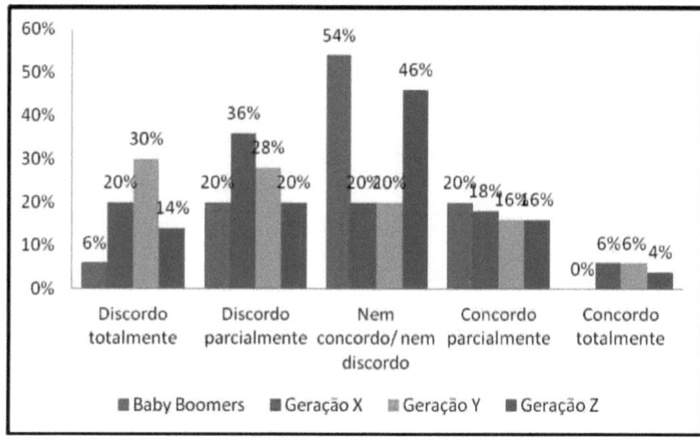

SOURCE: Research data (2018).

It can be seen that there were diverse results, but the neither agree nor disagree requirement resulted in higher rates for the *Boomers and Y* generations. This shows that these generations can feel compromised, depending on the technological change that may occur at work or in their personal daily lives.

In the case of totally disagree and partially disagree, the results were similar, with around 30% of those surveyed from all generations disagreeing. This shows that these people are afraid of challenges and feel committed to change, i.e. afraid of not learning how to use new technologies. And with indices below (20%) of both generations, the requirements were partially agree and totally agree. This shows that a small proportion of those surveyed are not intimidated by technological changes.

6.3.3 Resistance level

With regard to resistance, how do the generations feel that technological change is a threat? The following data was analyzed (Graph 13):

GRAPH 13: Feeling threatened by technological change.

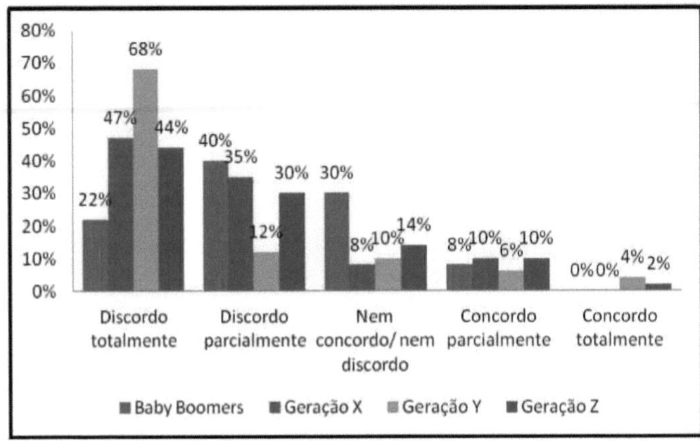

SOURCE: Research data (2018).

According to the data in Graph 13, in the totally disagree requirement, the Y generation scored highly (68%), the X generation (47%), the Z generation (44%) and the *Boomers* (22%), but in the partially disagree requirement, the *Boomers* scored the highest. This shows that even with different age groups and difficulties in handling them, the generations don't feel that technological changes are a threat, since they use them in their daily lives.

With regard to the feeling that technological change is detrimental to the daily lives of the generations, the following data is shown in Graph 14:

GRAPH 14: Feeling that technological changes are detrimental to everyday life.

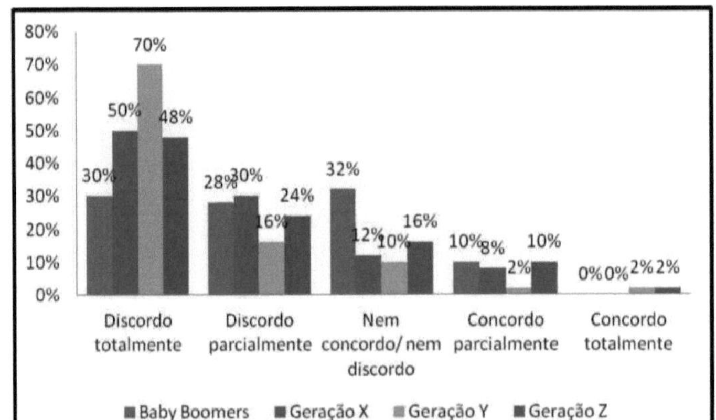

SOURCE: Research data (2018).

In view of the data, it can be seen that in the requirements of totally disagree and partially disagree, the majority of those surveyed, especially Generation Y, totally disagree that technological changes are detrimental to their daily lives, since technologies were designed to solve problems and help with day-to-day activities. This proves even more with the data obtained in Graph 8 that the generations agree that technologies are a way of making everyday life more practical.

Regarding rejection when a technological change occurs, the generations involve doing something they don't like, they perform the tasks slowly. The following data is shown in Graph 15:

GRAPH 15: Rejection of technological change in activities.

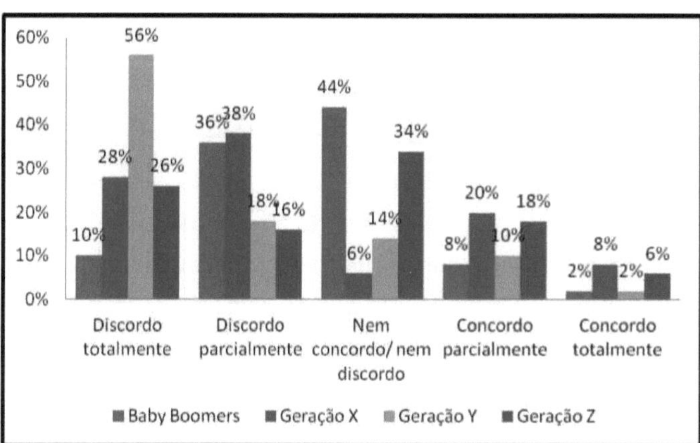

SOURCE: Research data (2018).

Given the data obtained, there was a diversity in the opinions of the generations, with Generation Y (56%) totally disagreeing with the requirement, i.e. they don't care about the technological change in their daily lives, they want to carry out their activities normally. With regard to the partial disagreement requirement, Generation X (38%) and the *Boomers* (36%), it can be seen that depending on the change they have to carry it out, i.e. if they don't like the change or are already used to the current one.

In the neither agree nor disagree requirement, the *Baby Boomer* generation (44%) and the Z generation (34%). This means that even if they don't like it, they carry out the tasks normally, and the requirements I agree partially and totally had low indices. A small minority agree that if they don't like it, they carry out the activities slowly, i.e. they are people who have reservations about entering into new experiences.

Graph 16 shows the generational data on whether they lack interest in carrying out activities that result in technological change:

GRAPH 16: Lack of interest in carrying out activities that result in technological changes.

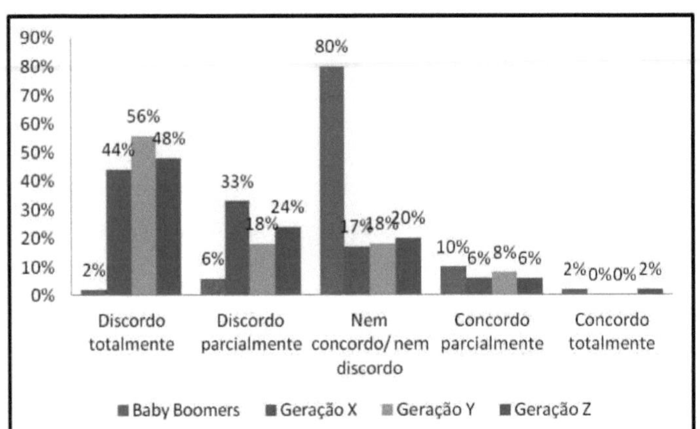

SOURCE: Research data (2018).

According to the data on neither agree nor disagree, the *Baby Boomer* generation obtained a higher index than the other generations with (80%) of those surveyed. This result can be compared with Graph 11, which showed that the *Boomers* were indifferent to experimenting with new technologies. Graph 16 shows that they are not interested in carrying out activities that will result in changes, which only makes this new issue more important.

For Generations X, Y and Z, the majority of those surveyed said they totally disagreed. Considerably, these generations like and use information systems and are interested in helping to improve technological changes. The other requirements scored low, i.e. only a small minority are not interested.

6.4 TECHNOLOGY USABILITY

The graphs below show questions that verified the ease of learning and satisfaction of the *Baby Boomer*, X, Y and Z generations in relation to the use of information systems. The results measured how the generations behave in their use, whether there are any difficulties and whether they feel confident in handling the technologies.

6.4.1 Ease of learning between generations

In relation to the ease of use of the technologies, the survey asked whether the generations found them easy to use. The following data was obtained (Graph 17):

GRAPH 17: Ease of handling technology.

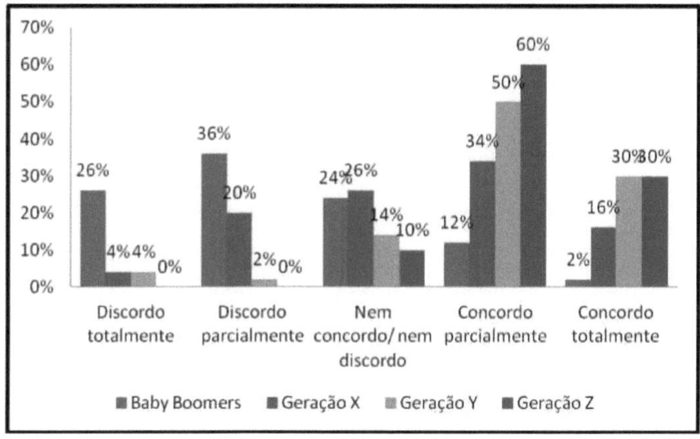

SOURCE: Research data (2018).

In view of the data, Generations Z, Y and X had the highest rates of partially agreeing that technology is easy to use. Considerably, as they were born using technology and depending on the equipment, most of them don't find it difficult, which supports the author Oliveira (2015) already mentioned in the study that these Generation Y young people have a faster learning capacity. With regard to the partially disagree and totally disagree requirements, the *Baby Boomers* obtained the highest scores, as they were born earlier than the other generations and had no interest in or financial conditions for requesting information systems, they find it more difficult to use technology.

With regard to the need for instruction in order to be able to handle any technological equipment, the survey asked whether the generations would need such instruction in order to use the technologies. This resulted in the following data (Graph 18):

GRAPH 18: Generations need instruction in handling equipment.

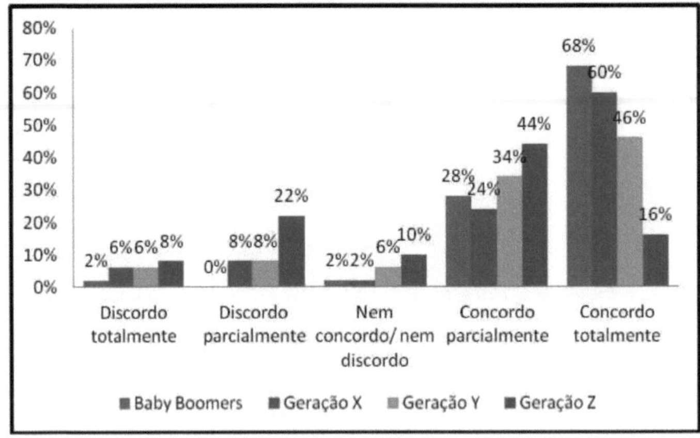

SOURCE: Research data (2018).

The requirements I partially agree and totally agree resulted in the highest rates with all generations, especially the *Boomers* and Xs. Considerably both generations need support to handle the technologies, with only a minority disagreeing that they don't need instruction.

Graph 19 shows the result of the opinion of each person surveyed in relation to other people who would learn to use technology quickly:

GRAPH 19: Opinion on learning quickly.

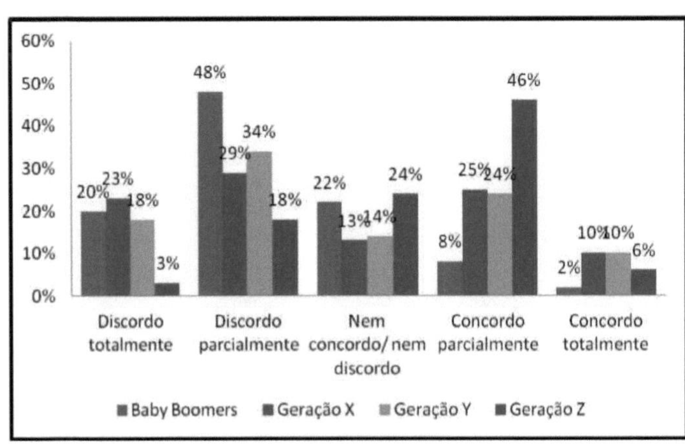

SOURCE: Research data (2018).

The opinions on this question were quite diverse, with *Baby Boomers* partially disagreeing (48%), Y (34%), X (29%) and Z (18%). This shows that they don't think that most people would easily learn how to use technology. The *Baby Boomer* generation had the highest score for this requirement, since they have difficulties handling it, they imagine that the other generations also have difficulties.

Generation Z (46%) of those surveyed thought that people would learn easily. Since they were born into the technological age, they imagine that other generations can learn as easily as they can handle equipment.

With regard to learning, the survey looked at whether the generations had to learn a number of things before they could continue to use the technologies and obtained the following data (Graph 20):

GRAPH 20: Learning needs for continued use.

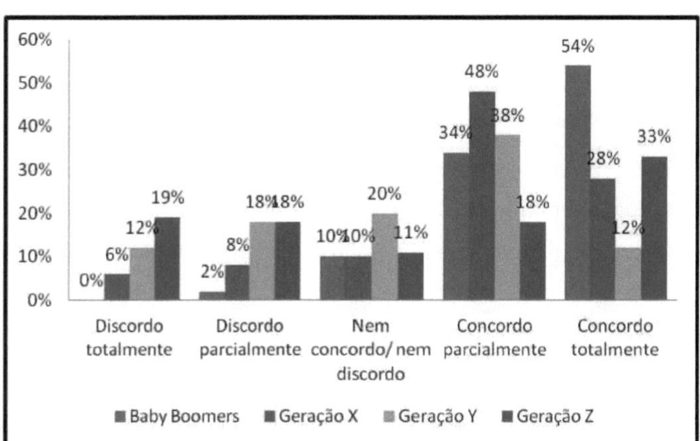

SOURCE: Research data (2018).

The results obtained showed a high score for the *Baby Boomer* generation in the totally agree requirement. This is probably because it has already been proven in the previous questions that this generation has difficulty handling equipment, which is why 54% of those surveyed had this need.

In the partially agree requirement, generations X (48%), *Boomers (*38%) and *Boomers* (34%), even though these generations found the usability of some equipment easy, as they were asked in Graph 17, needed other guidance on how to use it. There was a low score for the other requirements.

6.4.2 Satisfaction with the use of technology by generations

With regard to satisfaction with the use of information technology, the question was whether the generations would like to use technology frequently. The data from the analysis is shown in Graph 21:

GRAPH 21: Satisfaction with using technology frequently.

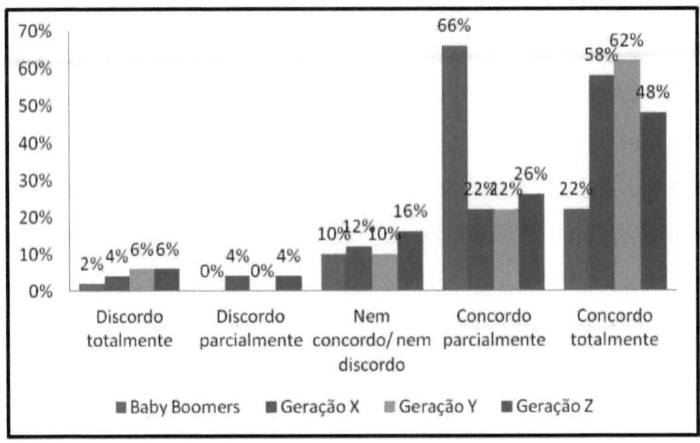

SOURCE: Research data (2018).

According to the data, the *Baby Boomer* generation stood out from the other generations in the partial agreement requirement. Despite all their fears and difficulties, this generation likes technology for its practicality and therefore prefers to use it frequently. Generations X, Ye and Z scored highly on the totally agree requirement, showing that technology is essential in everyday activities.

With regard to the question of confidence, the generations in handling technological equipment showed the following results (Graph 22):

CHART 22: Confidence in handling technology.

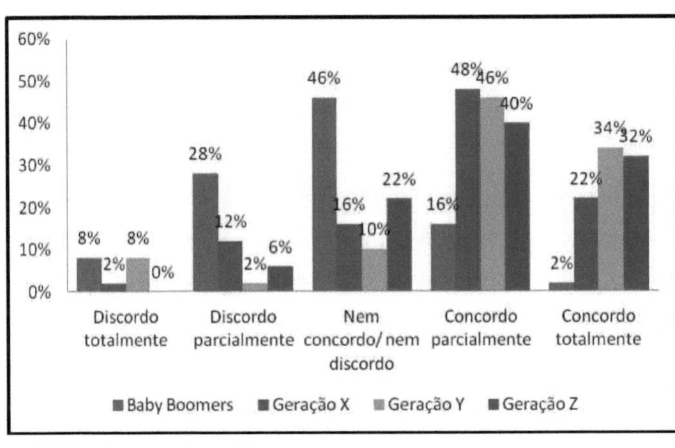

SOURCE: Research data (2018).

In view of the results, it can be seen that (28%) of those surveyed from the *Baby Boomer* generation partially disagree, showing that they do not feel confident in handling some technological equipment. The *Boomers* scored (46%) on the neither agree nor disagree requirement, probably because they are interested in handling it frequently, they prefer to remain indifferent on this requirement.

Generations X, Y and Z scored the highest in the partially agree and totally agree requirements. Considerably as they already use information systems in their daily lives and are not afraid of new technological experiences, they feel very confident in using this equipment.

7 FINAL CONSIDERATIONS

Considering that this study sought to analyze the evolution in the use of information systems and resistance to change by the *Baby Boomer*, X, Y and Z generations. Which was answered from the premise of the proposed objectives that were met, which were: to verify the evolution of the information systems available in each generation, to analyze the development of Information Technology and the adaptation of the generations in its use over time, and to identify the behavior of the different generations and their contributions to the use of information systems.

We looked at the evolution of the information systems available to each generation and analyzed their development over time, observing the differences between the generations in terms of their use of the technologies and information systems available. The *Baby Boomer* generation behaved differently from the others, although according to the theory they witnessed the birth of the computer, color TV and the radio and telephone. The *Boomers* indicated in the survey that most of them do not use *smartphones* or social networks, but rather television, cell phones and the radio, which has been forgotten by all generations. They find it difficult to handle technological equipment, but would like to use it frequently because of the practicality of the information, which indicates their participation in the transition and evolution of technologies over time.

Generation X, the daughter of the *Baby Boomers,* presents a balance and at the same time a divide between the characteristics presented by the generations. It resembles the *Baby Boomer generation*, as if its roots came from there, but its adaptation and acceptance of technologies and systems are similar to generations Y and Z, who were born in the wake of globalization and the constant changes and evolutions of the new era. In the survey, they showed that they use smartphones and computers more, but for work, unlike Y and Z, who indicated entertainment and communication. Given their age, most of them are in the professional world and are considerably the parents of generations Y and Z, so this use of technology differs from other generations.

Generation Y, individuals who were born during the evolution of the internet and already had easy access to information systems, showed in the survey that the majority use *smartphones* and that landlines were a technology of the past. This generation is looking for a higher professional position, which is why their interest in schooling is greater than that of previous generations and they are always training, and one of their characteristics, according to the theory cited in the research, is that they are quick and easy to learn.

Generation Z are today's young people, individuals with strong personalities who fight for freedom of expression and sexual differences. Young people who were born into the age of technology, similar in their behavior to Generation Y, but who use information systems to connect to social networks for fun and to communicate with each other. They don't find it difficult to adapt to new technological experiences, nor are they intimidated by their usability. Generations Y and Z showed greater familiarity with the new era of technology since they said they use technological equipment, making use of its facilities and its contributions during their daily lives.

In the light of the above, it can be concluded that the four generations that were analyzed, the first was resistant to change, but curious about experimenting with it; the second was a generation of overcoming and effort; and the last two, so far similar according to the data presented in the survey, were multitasking generations, used to and familiar with technology, unafraid of new experiences and in search of novelties and the facilities that advances can provide. In this study we have seen the evolution of information systems over the generations, from the *Boomers* with computers to the current Z generation with *smartphones,* and how they have reacted to the usability of technological information equipment.

It's worth noting that the research had some sampling limitations: the Generation Z sample profile showed a greater number of females than males, and males may have different behavior and opinions. The questionnaire was not selected randomly, but for convenience.

In addition, the profile of the sample also showed that the majority are single, people who may have fewer responsibilities and can spend more time using information systems. It is also worth mentioning that as the respondents were from different age groups, they may have misinterpreted the statements. This could be considered a limitation of this study.

It is also recommended that future research be carried out within the organizational context, in which the behavior of the generations at work can be analyzed, measuring the level of the relationship between technology and its users to see who has the most ability and least resistance to change in the development of the assigned activities. Another possibility would be to analyze each generation gender by gender, thus investigating the characteristics of the generations in greater depth in order to understand the reasons for their attitudes and attitudes towards technologies and systems. Finally, research could be carried out into the characteristics of the generations and the leadership profiles that exist in organizations.

8 REFERENCES

ABREU, KCK; SILVA, RS **História e Tecnologia da Televisao.** Available at: < http://www.bocc.ubi.pt/pag/abreu-silva-historia-e-tecnologias-da-televisao.pdf >. Accessed on September 20, 2017.

ANAUATE, C. **Geraçâo Z em Debate.** Available at: < https://www.edelman.com.br/white-paper/geracao-z-em-debate/> . Accessed on October 5 , 2017.

BALASSIANO, M. Relatório do projeto de pesquisa: electronic jogos eo mercado de trabalho. FAERJ, Rio de Janeiro, 2009.

BARBIERI, Luis Fernando. **Entenda os clientes da geraçao Baby Boomer** .

Available em :< http://revistastaoenegocios.uol.com . br/reportagens/entenda-os- clientes-da-geracao-baby-boomer/2559/> . Accessed: October 10, 2017.

BORTOLOTTI, SLV **Resistance to organizational change :** evaluation measure through the theory of responding to an item. PhD theses - Graduate Program in Production Engineering at the Federal University of Santa Catarina, 2010 .

BROOKE, J. SUS **- A quick and dirty usability scale.** Available at: < http://www.usabilitynet.org/trump/documents/Suschapt.doc >. Accessed on: December 14 , 2017.

BUSINESS INSIDER. **7 hâbitos que deixam a geraçâo Y mais anxious e unproductive** . Available to:

<http://epocanegocios.globo.com/Vida/noticia/2016/09/7-habitos-que-deixam- geracao-y-mais-ansiosa-e-improdutiva.html >. Accessed on November 8, 2017.

CASTRO, Caroline Lemes. **Technical Report:** Baby Boomer Generation . Available at: < http://www.ebah.com.br/content/ABAAAevDsAJ/relatorio-tecnico-sobre- geracao-baby-boomer >. Accessed: October 10, 2017.

CENNAMO, Lucy.; GARDNER, Dianne. **Generational differences in work values, outcomes and person** - organization values fit. Journal of Managerial Psychology, Bingley, v. 23, n. 8, p. 891-906, 2008.

CERETTA, SB; FROEMMING, LM Geraçâo Z: comprehendo os hâbitos de consumo da geraçâo emergente. **RAUnP - Revista Eletrônica do Mestrado Profissional em Administração da Universidade Potiguar** , v. 3, n. 2, art. 2, p. 15 24, 2011. Available at: < http://www.spell.org.br/documentos/ver/1395/geracao-z-- comprehendo-os-habitos-de-consumo-da-geracao- emerging >. Accessed on November 8, 2017.

CIRIACO, D. **O que é a geraçâo z?** Available em: < https://www.tecmundo.com.br/curiosidade/2391-o-que-ea-geracao-z-.htm >.

Accessed on October 12, 2017.

COIMBRA, RGC, & SCHIKMANN, R. (2001). **To generate net. Campinas:** Anais Anpad.

COMAZZETTO, LR, PERRONE, CM, VASCONCELOS, SJL, GONÇALVES, J. **A Geraçâo Y no Mercado de Trabalho: um Estudo Comparativo entre Geraçôes** . Available to:

<http://www.scielo.br/scielo.php?script=sci_arttext&pid=S1414-98932016000100145 >. Accessed on November 8, 2017.

CHIUZI, Rafael. A legacy of the Babys Boomers. Available em: < http://www.focoemgeracoes.com.br/index.php/2012/07/31/o-legado-dos-baby-boomers/> . Accessed on: October 1, 2017.

E-COMMERCE.ORG. Penetraçao da Internet em alguns paises. Available em: < https://www.e-commerce.org.br/ >. Accessed on: September 25, 2017.

FERREIRA, A. P. A invention of radio: um importante instrumento no contexto da disseminação da informação e do entertainment. **Multiplos Olhares em Ciência da Informaçâo** , v.3, n.1, mar. 2003. Available em: < http://portaldeperiodicos.eci.ufmg.br/index.php/moci/article/viewFile/1967/1237 >. Accessed on September 20, 2017.

FLINK, R., FERREIRA, CN, HONORATO, GM, ARAUJO, JR, E PROENÇA, TS (2012). **Porque e como atrair e reter os professionnels da Geraçâo Y nas empresas. In IX Congresso Virtual Brasileiro de Administração** . Available at: < http://www.convibra.com.br/upload/paper/2012/34/2012_34_5195.pdf >. Accessed on November 8, 2017.

GIL, Antonio. How to develop research projects. 5th Ed. São Paulo: Atlas, 2010.

GLASS, Amy. **Understanding generational differences for competitive success** . Jenkintown (PENN): BRODY Professional Development, 2007.

HELABS. Digital ser na medida certa para attarar a geraçâo X. Disponivel em: < https://helabs.com/blog/ser-digital-na-medida-certa-para-atingir-a-geracao-x/ >. Accessed on October 15, 2017.

IBGE - **Brazilian Institute of Geography and Statistics** . Available at: <http://www.bbc.com/portuguese/noticias/2015/04/150429_divulgacao_pnad_ibg e_lgb> Accessed on September 20, 2017.

INFOMONEY. Quem são, como vivem eo que pensam os jovens da Geraçâo Z? Available em: < http://www.infomoney.com.br/onde-investir/noticia/1859637/quem-sao- como-vivem-que-pensam-jovens-geracao >. Accessed on October 12, 2017.

KOKOSKA, Stephen. Introduce statistics. Sao Paulo: LTC, 2005.

KULLOK, Eline. Discover why entender a geraçâo Z pode aumentar 100% de suas chances de sucesso. Available em: < http://empreendedorismoconsciente.com/geracao/> . Accessed on November 8, 2017.

LEMOS, A. Ciberespaço e tecnologias móveis: procesos de territorialização e deterritorialização na cibercultura. In: COMPÓS - Annual Meeting, 2006, Bauru-SP.

Electronic analysis . Bauru: Associação dos Programas de Pôs-Graduação em Comunicação, 2006.

LEVY, P. *Cyberculture* . São Paulo: Edição brasileira (1999). Ed. 34 Ltda.

LOMBARDIA, P. G. **Quem é a geraçâo Y?** HSM Management, n.70, p. 1-7. Set./out. 2008.

MARCONI, M; LAKATOS, Eva. Fundamental to the scientific methodology. 7th Ed. São Paulo: Atlas, 2010.

MARTINS, Fabio. Senhores ouvintes, no ar... a cidade eo radio. Belo Horizonte: C/Arte, 1999. 140p.

MATTOS, ACM Information systems: an executive visionary . 2 ed. São Paulo: Saraiva, 2010.

MESSINA, A. P. A História da Informatica. Available em: < https://www.tecdicas.com/33/a-historia-da-informatica> . Accessed on September 20, 2017.

NACONECZNY, Santos and Baggio. Anàlise do behavior da Geraçâo Y no ambiente de trabalho na Cooperativa Castrolandia na cidade de Castro, Paranà.

Available em:< http://www.agr.feis.unesp.br/pdf/geracao_y.pdf >. Accessed on October 10 , 2017.

NETO, PA História das comunicaçôes e das telecomunicaçôes. Available at: <http://www2.ee.ufpe.br/codec/Historia%20das%20comunicaes%20e%20das%2

0telecomunicaes_UPE.pdf >. Accessed on September 20, 2017.

OLIVEIRA, S. Geraçâo Y: 'O que os jovens mais precisam nesse momento é de mentores'. Available em: < http://epocanegocios.globo.com/Carreira/noticia/2015/05/geracao-yo-que-os-jovens- mais-precisam-nesse-momento-e-de-mentores.html >. Accessed on November 8, 2017.

PACHECO, Roberto CS; TAIT, TFC Information Technology : Evolution and Applications. Teoria e Evidência Econòmica, Passo Fundo, v. 8, n. 14, May, 2000.

PENEDO, S. **Information Technology - Brief History and Perspectives.**

Available at: < http://agapedobrasil.com.br/blog/2015/04/28/tecnologia-da- informacao-breve-historia-e-perspectivas/ >. Accessed on September 20, 2017.

PEREIRA, AM; RENT, BS; PEDROSA, CFO; LACERDA, CHC;

FRANCO, G.; LUIZ, S.; SILVEIRA, VJA A história da Apple Computer. **Magazine**

Pretexto , v. 7, n. 1, p. 11-24, 2006. Available : <

http://www.spell.org.br/documentos/ver/27042/a-historia-da-apple-computer> . Accessed on September 20, 2017.

ROBBINS, SP; JUDGE, TA; SOBRAL, F. **Organizational behavior: theory and practice no contexto brasileiro** . 14. Ed. Sao Paulo: Pearson Prentice Hall, 2010.

SAVIEL, J. **A geraçâo Y e os professionnels multitarefas - Limites e Riscos.** 30 Dec 2009. Available at: < http://www.administradores.com.br/artigos/negocios/a- geracao-ye-os-profissionais-multitarefas-limites-e-riscos/37233/ >. Accessed on October 30, 2017.

SERRANO, Daniel Portillo. **Generation Baby Boomers** . 27 Jun 2010. Available em: < http://www.portaldomarketing.com.br/Artigos/Geracao_Baby_Boomer.htm >.

Accessed: October 15, 2017.

SOUZA, RL **Contemporary Carreiras e Novas Geraçôes Produtivas** . Sao Paulo: Atlas, 2010.

SORDI, José Osvaldo, MEIRELES, Manuel. **Information Systems Administration:** an interactive approach. Sao Paulo: Saraiva, 2010.

SULLIVAN, Sherry E. *et al* . Using the kaleidoscope career model to examine generational differences in work attitudes. Career Development International Bingley, v. 14, n. 3, pp. 284-302, 2009.

TAPSCOTT, Don. **A hora da geraçâo digital:** como os jovens que cresceram using the internet estao mudando tudo, das empresas aos governos. Rio de Janeiro: Agir Negócios, 2010.

TNS, Brazil. **Como a diferencia de geraçôes muda os hâbitos no mundo online** .

Available at: < https://exame.abril.com.br/marketing/como-a-diferenca-de- geracoes-muda-os-habitos-no-mundo-online/ >. Accessed on November 8, 2017.

LOOK. **The world has 3.2 billion people connected to the internet** . Available at: < http://g1.globo.com/tecnologia/noticia/2015/05/mundo-tem-32-bilhoes-de-pessoas-conectadas-internet-diz-uit.html >. Accessed September 30, 2017.

VECCHI, Ana. **Entenda os clientes da geraçao Baby Boomer** . Available at:< http://revistastaoenegocios.uol.com.br/reportagens/entenda-os-clientes-da-geracao-baby-boomer/2559/> . Accessed: October 10, 2017.

VEIGA, FJA, **Evolution of Information Systems** . University of Coimbra. Available at: < https://student.dei.uc.pt/~fveiga/GSI/Evolucao_Sist_Inf.pdf> .

Accessed on October 5, 2017.

VESCOVI, RA, **Os comportamentos de cooperação e competição entre as geraçôes nos ambientes de trabalho.** FUCK IT. Available em: < http://www.fucape.br/_public/producao_cientifica/8/Disserta%C3%A7%C3%A3o%20 Renata%20Agostini%20Vescovi.pdf >. Accessed on October 15, 2017.

WEINGARTEN, RM Four generations, one workplace: a gen xy staff nurse's view of team building in the emergency department. Elsevier, v.35, p. 27-30, 2011.

APPENDICES

FEDERAL INSTITUTE OF EDUCATION, SCIENCE AND TECHNOLOGY OF PARAÍBA

JOÀO PESSOA CAMPUS

ACADEMIC UNIT OF MANAGEMENT AND BUSINESS

HIGHER EDUCATION PROGRAM IN MANAGEMENT

Dear respondent, this data collection instrument from the research on **THE USE OF INFORMATION SYSTEMS OVER TIME: An Analysis of the Baby Boomer, X, Y and Z Generations. The** main objective of a study with students from the Administration course is to "analyze the evolution of the use of information systems by the *Baby Boomer*, X, Y and Z generations". The results of this research will be presented in the final paper of the Bachelor's Degree in Administration at IFPB and may also be presented at academic events and/or published in scientific journals, always protecting the identity of the respondents.

PROFILE OF THE INTERVIEWEE

1. **Gender** ( ) Female ( ) Male

2. **Which generation are you?**

( ) Born between 1940 and 1959 (Baby Boomers)

( ) Born between 1960 and 1979 (Generation X)

( ) Born between 1980 and 1989 (Generation Y)

( ) Born between 1990 and the present day (Generation Z)

3. **Marital status**

( ) Single

( ) Married

( ) Divorced

( ) Widowed

4. **Education**

( ) No education

( ) Completed 1st grade or completed elementary school

( ) Completed high school or secondary education

( ) 3rd grade or higher education completed

( ) Completed postgraduate degree

5. Family Income

( ) Up to R$ 937.00

( ) From R$ 937.01 to R$ 1,874.00

( ) From R$ 1,874.01 to R$ 3,748.00

( ) From R$ 3,748.01 to R$ 4,685.00

( ) Above R$ 4,685.00

ANALYSIS OF TECHNOLOGICAL USE

6. Do you like technology?

( ) Yes ( ) No

7. how often do you find it difficult to handle technological equipment?

( ) Never ( ) Sometimes ( ) Always

8. Which of these information technologies do you use most often today?

( ) Landline

( ) Television

( ) Radio

( ) Computer

( ) Cell phone

( ) Smartphones

( ) Tablet

( ) Other. What is it? ____________________________

9. Most of the time, you use information technology to:

( ) Fun and leisure

( ) Work

( ) Studies

( ) Shopping

( ) Communication

( ) Other. What is it? ____________________________

10. How many hours a day (on average) do you spend connected to online social

networks?

( ) Maximum 1 hour per day

(   )Between 1 and 4 o'clock in the morning

(   )Between 4 and 6 hours per day

(   )Between 6 and 8 p.m. wake up

(   )Over 8 hoursordia

( ) Other. Please specify: ____________________

11. In this question there are a series of statements related to resistance to change in the use of technology by generations, I ask you to indicate how much you DISAGREE or AGREE with each of the questions. On the scale, 1 indicates total disagreement, 5 indicates total agreement, and the other values indicate intermediate levels of agreement, as shown in the table below:

Level of Technology Use
1 - Totally disagree (DT)
2 - Partially disagree (SD)
3 - Neither agree nor disagree (I)
4 - Partially agree (CP)
5 - Totally agree (CT)

Construct	Statements about the use of technology	DT	DP	I	CP	CT
Acceptance	1. I am able to adapt to technological changes when they occur.	1	2	3	4	5
	2. I actively cooperate to bring about change when it happens.	1	2	3	4	5
	3. I am more likely to accept a technological change when I receive information about it.	1	2	3	4	5
	4. I believe that technological changes are a way of acquiring more practicality in my daily life.	1	2	3	4	5
Indifference	5. When technological changes happen, I try to do only what is necessary.	1	2	3	4	5
	6. I prefer to remain indifferent to technological changes.	1	2	3	4	5
	7. I prefer to do the same things in my daily life, rather than try different things.	1	2	3	4	5
	8. If technological changes happen, I don't feel compromised.	1	2	3	4	5
Resistance	9. I feel that technological change is a threat.	1	2	3	4	5
	10. I feel that technological changes in my routine affect my daily life.	1	2	3	4	5
	11. If technological change means doing something I don't	1	2	3	4	5

	like, I do it slowly.					
	12. I'm not interested in carrying out activities that will result in technological change.	1	2	3	4	5

12. In this question there are a series of statements related to the usability of technology by generations, I ask you to indicate how much you DISAGREE or AGREE with each of the questions. On the scale, 1 indicates total disagreement, 5 indicates total agreement, and the other values indicate intermediate levels of agreement, as shown in the table below:

Level of Technology Use
1 - Totally disagree (DT)
2 - Partially disagree (SD)
3 - Neither agree nor disagree (I)
4 - Partially agree (CP)
5 - Totally agree (CT)

Construct	Statements about usability	DT	DP	I	CP	CT
Ease of learning	1. I found the technology easy to use.	1	2	3	4	5
	2. I believe I would need instructions (support) to use the technology.	1	2	3	4	5
	3. I would have thought that most people would learn to use technology quickly.	1	2	3	4	5
	4. I had to learn a lot of things before I could continue to use the technologies.	1	2	3	4	5
Satisfaction	5. I would like to use technology frequently.	1	2	3	4	5
	6. I felt very confident using the technologies.	1	2	3	4	5

yes

I want morebooks!

Buy your books fast and straightforward online - at one of world's fastest growing online book stores! Environmentally sound due to Print-on-Demand technologies.

Buy your books online at
www.morebooks.shop

Kaufen Sie Ihre Bücher schnell und unkompliziert online – auf einer der am schnellsten wachsenden Buchhandelsplattformen weltweit! Dank Print-On-Demand umwelt- und ressourcenschonend produzi ert.

Bücher schneller online kaufen
www.morebooks.shop

info@omniscriptum.com
www.omniscriptum.com

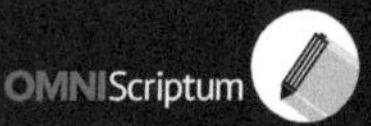

Printed by Books on Demand GmbH, Norderstedt / Germany